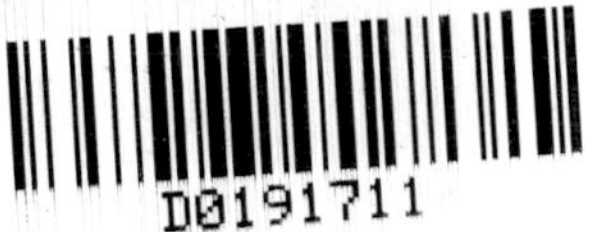

The Tiger That Isn't

'I have sat with Andrew Dilnot in many television studios and watched with awe as he eviscerates politicians who are trying to distort the figures to suit themselves. He is ruthless in exposing the lies that statistics can seem to support. In this witty and fascinating book he explains to us laymen how to make sense of numbers and how we can avoid having the wool pulled over our eyes. Invaluable.' – *David Dimbleby*

'...every journalist should get paid leave to read and reread *The Tiger That Isn't* until they've understood how they are being spun' – *New Scientist*

'This book is a valiant attempt to encourage healthy scepticism about statistics, against a culture in which both news producers and consumers like extreme possibilities more than likely ones' – *New Statesman*

'... a reliable guide to a treacherous subject, giving its readers the mental ammunition to make sense of official claims. That it manages to make them laugh at the same time is a rare and welcome feat' – *Economist*

'Very illuminating and comprehensible to even the mathematically-challenged' – *Thefirstpost.co.uk*

'An eye-opening lightning tour through the daily use and abuse of "killer facts": the way that statistics can beguile, distort and mislead ... This is essential reading for anyone interested in politics, economics or current affairs, and for whom "killer facts" have become part of daily rations. Mind your eye.' – *Scotland on Sunday*

The Tiger That Isn't

Seeing Through a World of Numbers

MICHAEL BLASTLAND

&

ANDREW DILNOT

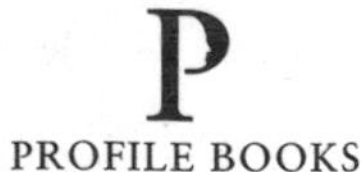

PROFILE BOOKS

First published in Great Britain in 2007 by
PROFILE BOOKS LTD
3A Exmouth House
Pine Street
Exmouth Market
London EC1R 0JH
www.profilebooks.com

10 9 8 7 6 5 4 3

Typeset in Palatino by MacGuru Ltd
info@macguru.org.uk

Printed and bound in Great Britain by
Bell & Bain Ltd, Glasgow

A CIP catalogue record for this book is available from the British Library.

ISBN 978 1 86197 839 4

Contents

To Catherine, Katey, Rosie,
Cait, Julia and Joe

1

Introduction

Numbers saturate the news, politics, life. For good or ill, they are today's pre-eminent public language – and those who speak it rule. Quick and cool, numbers often seem to have conquered fact.

But they are also hated, and often for the same reasons. They can bamboozle not enlighten, terrorise not guide, and all too easily end up abused and distrusted.

Potent but shifty, the role of numbers is frighteningly ambiguous. How can we see our way through them?

First, relax …

We all know more than we think we do. We have been beautifully conditioned to see through numbers, believe it or not, by our own experience. This is the radical premise of this book: it does not set out to pepper its readers with unfamiliar facts; instead, it shows how much they know already, and how to use that knowledge incisively.

Numbers can make sense of a world otherwise too vast and intricate to get into proportion. They have their limitations, no doubt, but are sometimes, for some tasks, unbeatable. That is, if used properly.

So this book is no litany of lies and damned lies. If its

authors – a journalist and an economist – really thought that badly of numbers and statistics we would have no business using so many, and nor should anyone else. Our aim is rather to bring them back to earth.

To achieve this, we do not depend only on uncovering the tricks of the trade: the multiple counting, dodgy graphs, sneaky start dates and funny scales – there have been exposés of that kind of mischief before – though there are gems in the stories that follow.

Nor do we rely on arcane statistical techniques, brilliant though those often are. Instead, wherever possible, we offer images from life – self, experience, prompts to the imagination – to show how to cut through to what matters. It is all there; all of us possess most of it already, this basic mastery of the principles that govern the way numbers work for the issues we follow and the causes we love or hate. It can be shared, we think, even by those who once found maths a cobwebbed mystery.

But simple does not mean trivial; it helps to answer imperative questions. Do we know what people earn and owe, who is rich and who poor? Is that government spending promise worth a dime? Who lives and who dies by government targets? Are those league tables honest? Do speed cameras save anyone? What about that survey of teenage offending, the 1 in 4 who do this, the 6 per cent increase in risk for women who do that, inflation, Iraqi war dead, HIV/Aids cases, the decline of fish stocks or hedgehogs, the threat of cancer, the pension time bomb, NHS budgets, waste and waiting times, third world debt, predictions of global warming? Hardly a subject is mentioned without measurements, quantifications, forecasts, rankings, statistics, targets, numbers of every variety; they are ubiquitous, and often disputed. If we are the least bit serious about any of them, we should attempt to get the numbers straight.

This means taking on lofty critics. Too many find it is easier to distrust numbers wholesale, affecting disdain, than get to grips with them. When a well-known writer explained to us that he had heard quite enough numbers, thank you – he didn't understand them and didn't see why he should – his objection seemed to us to mask fear. Jealous of his prejudices or the few scraps of numerical litter he already possessed, he turned up his nose at evidence in case it proved inconvenient. Everyone pays for this attitude in bad policy, bad government, gobbledegook news, and it ends in lost chances and screwed-up lives.

Another dragon better slain is the attitude that, if numbers cannot deliver the whole truth straight off, they are all just opinion. That damns them with unreasonable expectation. One of the few things we say with certainty is that some of the numbers in this book are wrong. Those who expect certainty might as well leave real life behind. Everyone is making their way precariously through the world of numbers, no single number offers complete and instant enlightenment, life is not like that and numbers won't be either.

Still others blame statistical bean counters for a kind of crass reductionism, and think they, with subtlety and sensitivity, know better. There is sometimes something in this, but just as often it is the other way round. Most statisticians know the limits of any human attempt to capture life in data – they have tried, after all. Statistics is far from the dry collection of facts; it is the science of making what subtle sense of the facts we can. No science could be more necessary, and those who do it are often detectives of quiet ingenuity. It is others, snatching at numbers, brash or over-confident, who are more naively out of touch.

So we should shun the extremes of cynicism or fear, on the one hand, and number idolatry on the other, and get on with doing what we can. And we can do a great deal.

Most of what is here is already used and understood in some part of their lives by almost everyone; we all apply the principles, we already understand the ideas. Everyone recognises, for example, the folly of mistaking one big wave for a rising tide and, since we can do that, perhaps to our surprise, we can unravel arguments about whether speed cameras really save lives or cut accidents. In life, we would see – of course we would see – the way that falling rice scatters and, because we can see it, we can also make simple sense of the numbers behind cancer clusters. We know the vibrancy of the colours of the rainbow and we know what we would lack if we combined them to form a bland white band in the sky. Knowing this can, as we will see, show us what an average can conceal and what it can illuminate. Many know from ready experience what it costs to buy childcare, and so they can know whether government spending in this area is big or small. We are, each one of us, the obvious and ideal measure of the policies aimed at us. These things we know. And each can be a model for the way numbers work. All we seek to do is reconnect what anyone can know with what now seems mysterious, reconnect numbers with images and experience from life, such that, if we have done our job, what once was baffling or intimidating will become transparent.

What follows will not be found in a textbook: even the choice and arrangement of subjects would look odd to a specialist, let alone the way they are presented. Good. This is a book from the point of view of the consumer of numbers. It is short and to the point. Each chapter starts with what we see as the nub of the matter: a principle, or a vivid image. Wipe the mental slate clean of anxiety or fuzziness and inscribe instead these ideas, keep each motif in mind while reading, see how they work in practice from the stories we tell. In this way we hope to light the path towards clarity and confidence.

The alignment of power and abuse is not unique to numbers, but it is just possible that it could be uniquely challenged, and the powerless become powerful. Here's how.

2

Size: Make It Personal

Simplify numbers and they become clear; clarify numbers and you stand apart with rare authority. So begin in the simplest way possible, with a question whose wide-eyed innocence defies belief:

'Is that a big *number?'*

Do not be put off by its apparent naivety. The question is no joke. It may sound trivial, but it captures the most underrated and entrenched problem with the way numbers are produced and consumed. Zeros on the end of a number are often flaunted with bravado, to impress or alarm, but on their own mean nothing. Political animals especially fear the size question, since their dirty secret is that they seldom know the answer. A keen sense of proportion is the first skill – and the one most bizarrely neglected – for all those interested in any measure of what's going on.

Fortunately, everyone already possesses the perfect unit of human proportion: their own self.

In November 2005 a front-page story in the *Daily Telegraph* reported government plans to raise the age of retirement for men from 65 to 67. If enacted, the paper said, one in five who would formerly have survived long enough to collect a pension would now die before receiving a penny. Hundreds of thousands would be denied by two cruel years.

1 in 5. Is that a *big* number?

In 1997 the Labour government said it would spend an extra £300m over five years to create a million new childcare places.

£300m. Is that a *big* number?

In 2006 the National Health Service was found to be heading for a budget deficit of nearly £1bn.

£1 billion. Is that a *big* number?

The answer to the first question is yes, 1 in 5 men aged 65 is a catastrophic number to die in two years, a number to strike terror into every 65 year old, a number so grotesquely enormous that someone at the *Telegraph* should surely have asked themselves: could it be true? Perhaps, if the plague returned, otherwise it doesn't take much thinking to see that the report was ridiculous. But it is a simple sort of thinking even smart journalists often do not do.

According to National Statistics (www.statistics.gov.uk), about 4 per cent of 65-year-old men die in the following two years, not 20 per cent. About 20 per cent of those born do indeed die before they reach the age of 67, but not *between* the ages of 65 and 67. Misreading a number in a table, as the journalists seem to have done, is forgivable; failure to ask whether the figure makes the sort of sense they could see every day with their own eyes, is less so.

Next, Labour's £300m for childcare. Here, no one involved in the public argument, neither media nor politicians, seemed to doubt its vastness. The only terms in which the opposition challenged the policy were over the wisdom of blowing so much public money on a meddlesome idea.

So is £300m to provide a million places a big number? Share

it out and it equals £300 per place. Divide it by five to find its worth in any one year (remember, it was spread over five years), and you are left with £60 per year. Spread that across 52 weeks of the year and it leaves £1.15 per week. Could you find childcare for £1.15 a week? In parts of rural China, maybe.

Britain's entire political and media classes discussed the policy as if you could. Does the public debate really not know what 'big' is? Apparently not, nor does it seem to care that it doesn't know. When we asked the head of one of Britain's largest news organisations why journalists had not spotted the absurdity, he acknowledged there was an absurdity to spot, but said he wasn't sure that was their job. For the rest of us, to outshine all this is preposterously easy. For a start, next time someone uses a number, do not assume they have asked themselves even the simplest question. Can such an absurdly simple question be the key to numbers and the policies reliant on them? Often, it can.

The third example, the £1bn NHS deficit, was roundly condemned as a mark of crisis and mismanagement, perhaps the beginning of the end for the last great throw of taxpayers' money to prove that a state system worked. But, was it a *big* number?

The projected deficit had fallen to about £800m at the time of writing, or about 1 per cent of the health service budget. But if that makes the NHS bad, what of the rest of government? The average Treasury error when forecasting the government budget deficit one year ahead is 2 per cent of total government spending. In other words, the NHS, in one of its darkest hours, performed about twice as well against its budgetary target as the government as a whole. There are few large businesses that would think hitting a financial target to within 1 per cent anything other than management of magical precision. They would laugh at the thought of drastic remedial action over so small a sum, knowing that to achieve better would require luck or a clairvoyant. NHS spending is equivalent to about £1,600 per head, per year (in 2007), of which 1 per cent is £16, or less than the cost of one visit to a GP (about £18). No doubt there was mismanagement in the NHS in 2006 and, within the total deficit, there were big variations between the individual NHS trusts, but given the immense size of that organisation, was it of crisis proportions, dooming the entirety?

Each time a reporter or politician turns emphatic and earnest, propped up on the rock of hard numbers like the bar of the local, telling of thousands, or millions, or billions of this or that, spent, done, cut, lost, up, down, affected, improved, added, saved … it is worth asking, in all innocence: 'is that a *big* number?'

In a famous cartoon mocking old-style Soviet central planning and target setting, the caption told of celebrations after a year of record nail production by the worker heroes of the Soviet economy, and the picture showed the entire year's glorious output: a single gigantic nail; big, but big enough for what? Claims of size are a con unless they are squared up to their purpose, but it is a con repeatedly fallen for. So it is often true that the most important question to ask about a number – and you would be amazed how infrequently people do – is the simplest.

In this chapter, six will be big and a trillion won't, without resorting to astronomy for our examples, but taking them from ordinary human experience. The human scale is what is often forgotten when size gets out of hand, and yet this is the surest tool for making numbers meaningful. Nor is there anything the least bit difficult about it. For it is also a scale with which we all come personally equipped.

For example, we asked people to tell us from a list of multiple-choice answers roughly how much the government spent on the health service in the UK that year (2005). The options ranged from £7 million to £7 trillion; so did people's answers. Since being wrong meant under- or overestimating the true figure by a factor of at least 10 and up to 10,000, it is depressing how wrong many were. The correct figure at that time was about £70 billion (70,000,000,000).

Some people find anything with an '-illion' on the end incomprehensible; you sense the mental fuses blowing at anything above the size of a typical annual salary or the more everyday mortgage. A total NHS budget of £7m is about the price of a large house in certain exclusive parts of London (and so somewhat unlikely to stretch to a new hospital), or equivalent to health spending per head of about 12p a year. How much healthcare could you buy for 12p a year? A total NHS budget of £7 trillion would have been more than six times as big as the whole economy, or about £120,000 of NHS spending a year for every single member of the population. When we told people that their guess of £7m implied annual health spending per head of about 12p, some opted instead for a radical increase to ... £70m (or £1.20p per head). How many trips to the GP would that pay for? How many heart transplants? These people have forgotten how big they are. Put aside economics, all that is needed here is a sense each of us already has, if only we would use it: a sense of our own proportions.

Millions, billions ... if they all sound like the same thing, a big blurred thing on the evening news, it is perhaps because they lack a sense of relative size that works on a human scale. So one useful trick is to imagine those numbers instead as seconds. A million seconds is about 11.5 days. A billion seconds is nearly 32 years.

If 300m can be pitifully small, and 1 in 5 outrageously big, how do we know what's big and what's small? (One of these is a number and the other a ratio, but both measure quantity.)

The first point to get out of the way is that the amount of zeros on the end of a number gets us nowhere; that much will be painfully obvious to many, even if some choose for their own reasons to ignore it. Immunity to being impressed by the words billion or million is a precondition for making sense of numbers in public argument. In politics and economics, almost all numbers come trailing zeros for the simple reason that there is a teeming mass of us and oodles of money in the economy of an entire country that produces more than £1,000,000,000,000 annually, as the UK does, with a population of 60,000,000 people. Plentiful zeros come with the territory. Our default position should be that no number, not a single one of them, regardless of its bluster, is either big or small until we know more about it. A recent (January 2007) much-hyped announcement to spend – gosh – £10m 'to boost singing and music education' in primary schools needs at least this much context: that there are around 10 million schoolchildren, roughly half of them in primary schools. So, £10m for 5 million children, per child, the boost would pay for what, exactly?

All this suggests an easy solution, already hinted at; that the best way to see through a number is to share it out and make it properly our own. Throwing away the mental shortcut – 'lots of zeros = big' – forces us to do a small calculation: divide the big number by all the people it is supposed to affect. Often that makes it strangely humble and manageable, cutting it to a size, after all, that would mean something to any individual, so that anyone can ask: 'Is it still big if I look at my share of it?'

Well, could you buy childcare for £1.15 a week? That is a judgement you can easily make. Will £300m pay for a million extra childcare places for five years? That sounds harder, though it is much the same question. Making the hard question easy is not remotely technically difficult; it is a matter mainly of having the confidence or imagination to do it.

Maybe on hearing numbers relating to the whole nation, some fail to make the personal adjustment properly: here's little me on average earnings and there's £300m. But it is not all yours. To make a number personal, it has to be converted to a personal scale, not simply compared with the typical bank balance. The mistake is a little like watching the teacher arrive with a bumper bag of sweets for the class and failing to realise that it means only one each. Yet being taken in by a number with a bit of swagger is a mistake that is often made. The largest pie can still be too small if each person's share is a crumb.

A convenient number to help in this sort of calculation is 3.12bn (3,120,000,000), which is the UK population (60,000,000), multiplied by 52, the number of weeks in a year. This is about how much the government needs to spend in a year to equal £1 per person per week in the United Kingdom. Divide any public spending announcement by 3.12 billion to see its weekly worth if shared out between us.

Some readers will be aghast that any of this needs saying, but the need is transparent and those in authority most often to blame. Part of our contention is that salvation from much of the nonsense out there is often straightforward, sometimes laughably so, and the simpler the remedy the more scandalous the need. Of course, not all cases are so routine, not all numbers need sharing equally, for example, often being aimed more at some than others. That's another simple but important principle to which we will return. For now it serves to emphasise the point that we can make more sense of a number if we first share it out where it is due.

None of this is to encourage cynicism, it is about how to know better. And it is worth saying that innumeracy is not the same as dishonesty. Numbers do grow because people try to pull the wool over our eyes, sure enough, but also because they are themselves muddled, or so eager for evidence for their cause that they forgo plausibility. Maybe the *Daily Telegraph* journalists, writing about dying pensioners, were so taken with the notion that Blair and company were beastly to old folk who had slaved for an honest crust, that they allowed this apparently satisfying notion to make mush of their numeracy.

This tendency of big brains to become addled over size is why the peddlers of numbers often know less about them than their audience, namely, not much; and why questions are always legitimate, however simple. Size matters. It is odd trying to persuade people to give it more attention, but this is attention of a neglected kind, where, instead of simply bowing to a number with apparent size and attitude, we insist on a human-scale tape measure.

Size is especially neglected in the special case when mixed with fear. Here, there is often no need even to claim any great magnitude; the simple existence of danger is enough, in any quantity.

To see this, and the kind of brainstorm that talks of toxicity or poison as if it means invariably fatal by the micron, think of jumping. Performed from the bottom step, it is seldom harmful; from the rooftop it is more serious, frequently as serious as harm can be; and without something soft to land on … the details are not necessary. It is clear that jumping from a height has variable consequences depending, well, on the height. Readers of this book asked to jump from a height would, we hope, not be alone in asking, 'How high?'

This humdrum principle – that harm depends strictly on how much of a given danger people are exposed to – is one that everyone applies unthinkingly umpteen times a day. Except to toxicity. Food and environmental health scares are a paranoia-laced anomaly, often reported and, presumably, often read with no sense of proportion whatsoever, headlined with statements the equivalent of 'Jumping fatal, says study'.

Let's labour the metaphor to bring out its relevance. Jumping from a height has variable consequences, depending on the dose. So what happens to this elementary principle when, as happened in 2005, someone tells us that cooked potatoes contain a toxic substance called acrylamide?

Acrylamide is used industrially – and also produced by a combination of sugars and amino acids during cooking – and known at some doses to cause cancer in animals and nerve damage in humans. The scare originated in Sweden in 1997 after cows drinking water heavily polluted with industrial acrylamide were found staggering around as if with BSE. Further Swedish research, the results of which were announced in 2002, found acrylamide in a wide variety of cooked foods.

On average, people are reckoned to consume less than 1 microgram of acrylamide per kilogram of body weight per day. That is 1/1000th of the dose associated with a small increase in cancer in rats. Some will consume more, though in epidemiological studies have seemed to get no more cancer than anyone else.

Salt is often added to potatoes, particularly chips or crisps, liberally. It is both essential for survival and also so toxic it can kill a baby with smaller quantities than may be contained in a salt cellar. About 3,750 milligrams per kilo of body weight is the accepted lethal dose (LD) for salt (the quantity that has been found to kill 50 per cent of rats in a test group, known as the LD50). For a 3kg baby, that equals about 11 grams, or a little more than two teaspoons. Does anyone report that two teaspoons of salt sprinkled on your chips flirts with death? What is said, for a variety of health reasons, is that we should keep a casual eye on the quantity we consume. Be sensible, is the unexcitable advice. Even a bit of arsenic in the diet is apparently essential for good health.

So why such absolutism about the quantities of acrylamide? Maybe the dual advantages of health paranoia and never having heard of the stuff before made panic more inviting. Had journalists thought about the numbers involved, it might have kept fear in proportion: the quantity of acrylamide equivalent to that associated with a small increased risk of cancer in rats turned out to require consumption of about 30kg of cooked potatoes (about a third to a half of the typical human body weight) every day for years.

When the headline reveals toxicity, the wise reader, aware that most things are toxic at some dose, asks, 'In what proportions?' Water keeps you alive, unless, as thirsty consumers of the drug ecstasy are often told, you drink too much, when it can cause hyponaetraemia (water-poisoning). Leah Betts, a schoolgirl, died after taking ecstasy then drinking 14 pints of water in 90 minutes.

That is why all toxicologists learn the same mantra from their first textbooks, that toxicity is in the dose. This does not mean all claims of a risk from toxicity can be ridiculed, simply that we should encourage the same test we apply elsewhere, the test of relevant human proportion. Keep asking the wonderful question, with yourself in mind: how big is it?

There are exceptions: peanut allergies for example can be

fatal at such small doses that it would be unhelpful to say the sufferer must quibble over the quantity. Otherwise, though treated like one, 'toxic' is not necessarily a synonym for 'panic!'.

By now, you might be all but convinced of the merit of our coy little question, and that long lines of zeros tell us nothing. So let's measure that assurance with a real monster of a number: £1,000,000,000,000 – or one trillion pounds, as we now call such a quantity of zeros, in accordance with international economic convention. Imagine this was what British people owed, the total of their debts. Do not be intimidated by the apparent size, go ahead and ask: Is that a *big* number?

Most newspapers thought so, when we reached that figure in 2006, and some splashed it across the front page. The answer is that it is highly debatable whether debts of £1 trillion today are large or not. It was the highest figure ever, true, since surpassed, but we can also say with supreme confidence that it will be surpassed again, and again, since records are all but inevitable in a growing economy. The reporting implied surprise; but the combination of inflation and a growing economy means that the total number of pounds in Britain rises by about 5 per cent a year. This doubles the number of pounds in the economy every fifteen years. When there are so many more pounds of everything, is it any surprise there are also more pounds of debt?

So put aside the manufactured sense of shock at this number and try pitching at it one of our innocent little questions: how is it shared out? Not evenly, is the unsurprising answer. Nor is it the prototypical shopaholic with a dozen maxed-out credit cards who accounts for any but a tiny part of the trillion. It is, in fact, the rich who owe overwhelmingly the biggest proportion of the debt, and always have, often in mortgages taken out to pay for houses that are growing in value, which means they also have increasing assets to repay their debts if they have to.

To see why debt is likely to be shared out like this, ask yourself how much you would be content to borrow prudently aged 16, compared with what you might find yourself manageably owing aged 42. So far as the sum has risen, does it represent a shocking descent into middle-aged profligacy? Or is larger debt mostly a reflection of larger borrowing power, linked to a rising ability to sustain higher repayments? The national picture is predictable, given the typical personal experience. Debt is a corrosive problem for those who cannot afford it, and there do genuinely seem to be more people in that position, but this has precious little to do with the trillion pounds, most of which is a sign not of penury, but wealth.

Another way of thinking about this is to say that your debts would be too big only if you could not repay them. But how many news reports of UK debt also report UK wealth? Quite absurdly, it almost never gets a mention, even though it would be one of everyone's first considerations. So trust what is known to be true for the individual, take the calculation anyone would make personally, and apply it to the national picture. This shows (see Figure 1) that personal wealth has been increasing over recent years steadily and substantially. In 1987 the total personal wealth of all the households in the UK added up to about four times as much as the annual income of the whole country. By 2005 wealth was six times as great as annual national income. Not poorer, but prodigiously richer, is the story of the last twenty or so years. This wealth is held in houses, pensions, shares, bank and building society accounts, and it is held very unequally, with the rich holding most of

Figure 1 **Wealth and debt in the UK**

As a multiple of GDP

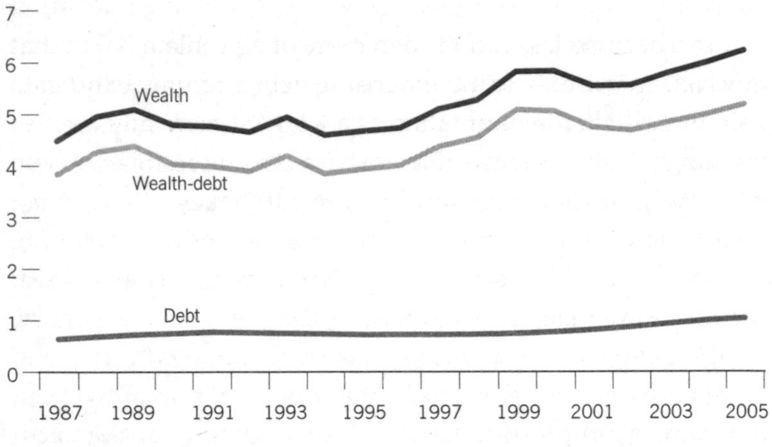

the wealth, as well as most of the debt. Some of the increase reflects house-price growth, some increases in share prices, but there is no denying that as the economy has grown, and incomes with it, so have savings and wealth. Debt has been rising, but wealth has been rising much more quickly.

Even if we look at the category of debt said to be running most seriously out of control – unsecured personal debt (which includes credit cards) – we find that as a proportion of what we have available to spend – household disposable income – this debt has remained steady for the last five years (before which it did go up – a bit). Once more, it is unequally distributed and more of a problem for some than others, but for precisely those reasons it makes no sense to use the total number as if it says anything whatsoever about those in direst financial straits.

How can we know from personal experience that the debt number may not be as monstrous as it looks? By thinking about our own life cycle and our own borrowing. We are happier to borrow more as we grow richer, and often do, but it tends to become less rather than more of a problem. With that much clear, it is easy to see that rising debt might well indicate rising wealth in the population as a whole. Essentially, we use the same method as before, we share out the number where it belongs, rather than according to what makes a good story. So we should share it mostly among the rich, far less among the poor. Just for good measure, £1 trillion of debt shared out across the population is a little less than £17,000 per person, compared to wealth, on average, of over £100,000.

It is not only debt in the UK that can more usefully be put into human proportions. In the summer of 2005 at the Gleneagles summit – held during the UK's presidency of the G8 group of countries – the Chancellor of the Exchequer Gordon Brown announced that $50bn of developing country debt would be written off. It sounds generous, and is important for developing countries, but how generous does it look from inside the G8? Is it a *big* number for us?

Fifty billion dollars was the stock of outstanding debt. By writing if off, what the G8 countries actually gave up were the annual repayments they would have received, often at preferential interest rates, equal to about $1.5bn a year. This was the real cost, substantially lower than the headline figure. Convert $1.5bn into pounds, and it seems smaller still, at around £800m at the then prevailing exchange rates. £800m a year still sounds a lot, but remember the exhortation to make numbers personal. The population of the G8, conveniently, is around 800m. So the Gleneagles deal will cost each person living in the G8 about £1 a year. And even that is an exaggeration, since much of the money will come out of government aid budgets that would have gone up anyway, but will now be diverted to pay for the Gleneagles deal, leaving the best estimate of how much more Gleneagles will cost each of us than had it never happened, as close to zero.

Let's look, finally, at the other end of the scale, to a number in single figures, a tiny number: six. You will be honing your scepticism by now, reluctant to take anyone's word or commit to an opinion, and wanting to know more of the context. That is as it should be. The six we have in mind is the celebrated six degrees of separation said to exist between any two people on the planet. So, is six a small number?

The idea was first proposed – according to Wikipedia – in 1922 in a short story titled 'Chains' by a Hungarian writer, Karinthy Frigyes. But it was an American sociologist, Stanley Milgram, who became most closely associated with the suggestion when, in 1967, he claimed to have demonstrated empirically that six steps was generally more than enough to connect any two people in the United States. He called this 'the small world phenomenon'.

Milgram recruited nearly 300 volunteers whom he called 'starters', asking each to send a gold-embossed passport-like package to a stranger, usually, but not always, in another city. They could use any intermediary they knew on first-name terms, who would then try to find another who could do the same, each time moving closer to the target. All they had to go on was the recipient's name, home town, occupation, and a few descriptive personal details, but no address. Milgram reported that 80 per cent of packages reached their target in no more than four steps and almost all in fewer than six.

The story became legendary. Then, in 2001, Judith Kleinfeld, a psychologist at the University of Alaska, became interested in the phenomenon and went to study Milgram's research notes. What she found, she says, was disconcerting: first, that an initial unreported study had a success rate of less than 5 per cent. Then, that in the main study more than 70 per cent of packages had never reached their intended destination, a failure rate that raises doubts, she says, about the whole claim. 'It *might* be true,' she told us, 'but would you say that the packages arrived in fewer than six steps when 70 per cent never arrived at all?' Furthermore, in criticism that bears more closely on our question, she notes that senders and recipients were of the same social class and above-average incomes, and all likely to be well connected.

So six may or may not be the right number, and the central claim in the experiment has never been satisfactorily replicated, but would it, if true, nevertheless be a small number? The point that connections are easier among similar people is a clue, and a prompt. This encourages us to think that not only the number of steps, but also the size of the steps matters. And if each step is giant, six will amount to an impossible magnitude.

Other studies, including more by Milgram himself, according to Judith Kleinfield's summary of his personal archives, found that where the packages crossed racial differences, the completion rate was 13 per cent, rising to 30 per cent in a study of starters and targets living in the same urban area. When, in another study, the step was from a low-income starter to a high-income target, the completion rate appears to have been zero. Connections were not nearly so easy across unexceptional social strata.

And as Judith Kleinfeld herself points out: 'A mother on welfare might be connected to the President of the United

States by a chain of fewer than six degrees: her caseworker might be on first name terms with her department head, who may know the Mayor of Chicago, who may know the President of the United States. But does this mean anything from the perspective of the welfare mother? … We are used to thinking of "six" as a small number, but in terms of spinning social worlds, in a practical sense, "six" may be a large number indeed.'

'Six' is usually small; a billion is usually large. But easy assumptions will not do when it comes to assessing size. We need to check them with relevant human proportion. Six steps to the President sounds quick, but let people try to take them. They can be an ocean apart and represent a world of difference. A billion pounds across the UK can be loose change, 32p each per week. We need to think, just a little, and to make sure the number has been properly converted to a human scale that recognises human experience. Only then, but to powerful effect, can we use that personal benchmark. The best prompt to thinking is to ask the question that at least checks our presumptions, simple-headed though it may sound: 'Is that a *big* number?'

3
Counting: Use Mushy Peas

Counting is easy when you don't have to count anything with it: 1, 2, 3 … it could go on for ever, precise, regular and blissfully abstract. This is how children learn to count in the nursery, at teacher's knee.

But for grown-ups putting counting to use in the wider world, it has to lose that innocence. The difference is between counting, which is easy, and counting something, *which is anything but. Some find this confusing, and struggle to put the childhood ideal behind them.*

But life is more complicated than a number. The first is multidimensional and shape-shifting, the other flat and precise. That is why, when counting something, *we risk torturing reality, squashing and mashing it into something else until it fits the numbers, when the process should be the other way round. And then, worst of all, the fact that all this was a struggle is forgotten. To avoid this mistake, and master counting in real life, renounce the memory of the classroom and follow a better guide: mushy peas.*

'Yob Britain! 1 in 4 teen boys is a criminal!' said headlines in January 2005. '1 in 4 teen boys claims they have done a robbery, burglary, assault or sold drugs.' As another newspaper put it: 'Welcome to Yob UK!'

A survey of teenage boys suggested Britain had nurtured a breed of thugs, thieves and pushers. Newspapers mourned parenting or civilisation, politicians held their heads in their hands in woe and the survey itself really did say that 1 in 4 teenage boys was a 'prolific or serious offender'.

A competent survey, which this was, asking boys what they'd been up to and then counting the answers, ought to be straightforward (assuming the boys told the truth). Counting, after all, is simple enough for the nursery, where one number follows another, each distinct and all units consistent: 1, 2, 3 …

The common, unconscious assumption is that this child's-play model of counting still applies, that it is a model with iron clarity in which numbers tick over like clockwork to reach their total. But when counting anything that matters in our social or political world, although we act as if the simple rules apply, they do not, they cannot, and to behave otherwise is to indulge a childish fantasy of orderliness in a world of windblown adult jumble.

It is, for a start, a fundamental of almost any statistic that, in order to produce it, something somewhere has been defined and identified. Never underestimate how much nuisance that small practical detail can cause.

First, it has to be agreed what to count. What is so hard about that? Take the laughably simple example of three sheep in a field:

What do we want to count?
Sheep.
How many are there?
Three.

But one is a lamb. Does that still count as a sheep? Or is it half a sheep? One is a pregnant ewe in advanced labour. Is that one sheep, or two, or one and a half? (We assume no multiple births.) So what is the total? Depending how the units are defined, the total could be 2, 2.5, 3, 3.5 or 4. For a small number, that's quite a spread; one answer is twice the size of another, and counting to four just became ridiculous. In the real world, it often does whenever the lazy use of numbers belittles the everyday muddle.

If a simple case is complicated, think what goes on in government statistics aiming to capture changes in the lives of millions of people, prices and decisions. Take the number out of work. When do we call someone unemployed? Must they be entirely unemployed or can they do some work and still qualify? How much? An hour a week, or two, or ten? Do they have to look actively for work or just be without it? If they do have to look, how hard? What if they work a bit, but they are not paid, as volunteers? Mrs Thatcher's Conservative governments notoriously changed the definition of 'unemployed' twenty-three times.

Numbers, pure and precise in abstract, lose precision in the real world. It is as if they are two different substances. In maths they seem hard, pristine and bright, neatly defined around the edges. In life, we do better to think of something murkier and softer. It is the difference, in one sense, between diamonds and mushy peas; hard to believe it is a difference we forget or neglect. Too often counting becomes an exercise in suppressing life's imprecision.

In that real world of soft edges, what was the definition used to identify those teenage boys as serious or prolific offenders? The largest category of offence by far was assault, and judged serious if it caused injury. Here is how it was defined by a survey question:

> Have you ever used force or violence on someone on purpose, for example by scratching, hitting, kicking or throwing things, which you think injured them in some way?

And then, deliciously:

> Please include your family and people you know as well as strangers.

Fifty eight per cent of assaults turn out to have been 'pushing' or 'grabbing'. Thirty six per cent were against siblings. So anyone who pushes a brother or sister six times leaving them no worse for wear is counted a 'prolific offender'. Push little brother or sister even once and bruise an arm and you are a 'serious offender', since the offence led to injury. Or, as the press put it, you are a yob, bracketed with drug dealers, burglars, murderers and every other extremity of juvenile psychopath.

Each time we count something, we define it; we say the things we count are sufficiently the same to be lumped together. But most of the important things we want to count are messy, like people, behaving in odd ways, subtly and not so subtly different. They do not stand still, they change, their circumstances vary hugely. So how do we nail them down in one category under one set of numbers? Using what definitions? Until this context is clear, how can it even be said what has actually been counted?

Behaviours that seem clear-cut and therefore easily countable at the extremes often blur into one another somewhere along a line of severity. Sister-shovers and knife-wielding maniacs differ, so why are they lumped together in the reporting of this survey? In determining a definition of what to count, the researchers looked to the law, whose long arm can indeed be brought to bear on pushing, shoving etc.; can be, but normally isn't. The wise copper on the beat who decides that the brothers with grazed elbows from No. 42 are squabbling kids, not criminals fit for prosecution, is making a human judgement with which formal definition struggles. Definition does not like discretion and so tempts us to count rigidly, as teacher taught, in this case by saying that behaviour is either illegal or not. But this fancifully optimistic classification, insisting on false clarity, serves only to produce a deceptive number. It is astonishing how often the '1 in 4' headline comes along for one social phenomenon or another. It is invariably far tidier than it ought to be, often to the point of absurdity. The paradox of this kind of neatness is that it obscures; it is a false clarity leading to a warped perspective, and an occupational hazard of paying insufficiently inquiring attention to the news, in numbers above all.

Some of the crime identified by the survey was certainly brutal but, given the definitions, what were the headline numbers from the survey really counting? Was there clear statistical evidence of Yob UK? Or did the survey rather reveal a remarkably bovine placidity in Britain's sons and brothers? After all, 75 per cent of all teenage boys claimed not to have pushed, grabbed, scratched or kicked as many as six times in the past year (what, not even in the school dinner queue?), nor to have done any of these things once in a way that caused even the mildest scrape. According to these definitions, the authors of this book must make their confession: both were serious or prolific offenders in their youth. Reader, your writers were yobs. But perhaps you too were one of us?

We make no argument about whether the behaviour of Britain's youth is better or worse than it used to be. We note examples of appalling behaviour like everyone else. We do argue that the 1 in 4 statistic – reported in a way that implied things were getting worse – was no evidence for it, being no evidence of anything very much. Numbers, claiming with confidence to have counted something breathtaking, mug our attention on every corner. This survey told us much of interest, if we could be bothered to read it properly, but the summary statistic, '1 in 4', counted for nothing.

It wouldn't sell, but a more accurate tabloid headline might have read: 'One in four boys, depending how they interpret our question, admits getting up to something or other that isn't nice, a bit thoughtless maybe, and is sometimes truly vicious and nasty; more than that, we can't really say on the evidence available to us.'

Let us assume the details are usually worked out properly, that we agree whether we mean sheep only or sheep *and* lambs, *and* we say so in advance, *and* we've decided on a clear way of telling the difference, and that what reports of the survey meant to say was that if you look at enough boys you will find a range from brutal criminal thug, through bully, all the way to occasionally unthinking or slightly boisterous lad, now and then, in the dinner queue, maybe. Less dramatic, less novel, and oh, all right, you knew it already; but at least that would be a little more accurate. Sticking a number onto trends in juvenile behaviour gives an air of false precision, an image of diamond fact, and in this case forces life into categories too narrow to contain it. Picture that offending more properly as a vast vat of mushy peas, think again about what single number can represent it, and see this massively difficult measurement for what it is. The fault here is not with numbers and the inevitable way that they must bully reality into some semblance of orderliness. It is with people, and their tendency to ignore that this compromise took place, while leaping to big conclusions.

Is this a mere tabloid extravagance? Not at all: it is commonplace, in policymaking circles as in the media.

When, amidst fears of a pensions crisis, the government-appointed Turner Commission published a preliminary report in 2005 on the dry business of pension reform, it said 40 per cent of the population was heading for 'inadequate' provision in retirement. With luck, your definitional muscles will now be flexing; what do they mean by 'inadequate'?

In reaching that 40 per cent figure – a shocking one – the Commission had said each to their own resources: either you have a pension or you don't, a hard-and-fast definition like the interpretation in the yob survey of the law on assault, in or out but nothing in between, covered or not, according to your own and only your own finances.

But perhaps you married young, perhaps you never had a paid job, perhaps your partner earned or still earns a sum equal to the GDP of a small state, perhaps you had been together for forty years, living in a château. All things considered, you might expect a comfortable sunset to life. But if you were fool enough in the Commission's eyes to have no savings or pension in your own name – shame on you – you were counted as lacking adequate provision for retirement.

This was a figure that mattered greatly to policymakers, since it seemed to suggest on an epic scale an irrational population in denial about the possibility of old age. Marriage – hardly an unreasonable detail – was one of many definitional question marks over that assessment. Soon afterwards another report, an independent one, recalculated the number it believed to be without adequate pensions, this time taking marriage into account. It came up with 11 per cent with inadequate savings instead of 40 per cent. In later reports by the Pensions Commission itself, the 40 per cent claim disappeared.

For the time being, we can establish a simple rule: if it has been counted, it has been defined, and that will almost always have meant using force to squeeze reality into boxes that don't fit. This is ancient knowledge, eternally neglected. In Book I of Aristotle's *Nichomachean Ethics* he writes, 'It is the mark of an educated man to look for precision in each class of things just so far as the nature of the subject admits.' But not more and that is the critical point. Always think of the limitations. Always ask: are the definitions diamond hard or mushy pea? Either way, am I content with the units they define? In short: 1, 2, 3 of what? And are there really 4, 5 and 6 of the same?

4

Chance: The Tiger that Isn't

We think we know what chance looks like, expecting the numbers she wears to be a mess, haphazard, jumbled. Not so, for chance has a genius for disguise and fools us time and again. Frequently and entirely by accident, she appears in numbers that seem significant, orderly, coordinated, or slip into a pattern. People feel an overwhelming temptation to find meaning in these spectral hints that there is more to what they see than chance alone, like zealous detectives over-alert to explanation, and to dismiss with scorn the real probability: '*it* couldn't *happen by chance!*'

Sometimes, though more seldom than we think, we are right. Often, we are suckered, and the apparent order is no order, the meaning no meaning, it merely resembles one. The upshot is that discovery after insight after revelation, all claiming to be dressed in the compelling evidence of numbers, will in fact have no such respectability. It was chance that draped them with illusion. Experience offers us innumerable examples. We should take them to heart. In numbers, always suspect that sly hand.

Who was it, furtive, destructive, vengeful in the darkness? At around midnight on bonfire night – 5 November 2003 – on the outskirts of the village of Wishaw in the West

Midlands, someone had crime in mind, convinced the cause was just.

Whoever it was – and no one has ever been charged – he, she, or they came with rope and haulage equipment. A few minutes later, on the narrow stretch between a riding stables and a field, the 10-year-old, 23-metre tall mobile-phone mast on the outskirts of the village was first quietly unbolted, then brought crashing down. The signal from the mast ceased at 12.30 a.m. precisely. Police found no witnesses.

By morning, protestors surrounded the upended mast and refused to allow T Mobile, its owners, to take it away or replace it. A solicitor representing the protesters told the landowners they would not be permitted access because that meant crossing someone else's property. The protest quickly became a round-the-clock vigil with both sides paying private security companies to patrol the boundary.

The villagers' earnest objection had a despairing motivation. Since the mast had gone up, among the twenty households within 500 metres, there had been nine cases of cancer. In their hearts, the reason seemed obvious. They were, they believed then and still believe now, a cancer cluster. How could such a thing happen by chance? How could so many cases in one place be explained except through the malign effect of powerful signals from that mast?

The villagers of Wishaw might be right. The mast has not been replaced and the strength of local feeling makes that unlikely, not now, perhaps not ever. And if there were a sudden increase in crime in Wishaw such that nine out of twenty households were burgled, they would probably be right to suspect a single cause. When two things happen at the same time, they are often related.

But not always, and if the villagers are wrong, the reason

has to do with the strange ways of chance in large and complex systems. *If* they are wrong, it is most likely explained as a result of their inability – an inability most of us share – to accept that apparently unusual events happening simultaneously do not necessarily share the same cause, and that unusual patterns of numbers in life, including the incidence of illness, are not at all unusual, not necessarily due to some guiding force or single obvious culprit, but callously routine, normal and sadly to be expected.

To see why, stand on the carpet – but choose one with a pile that is not too deep (you might in any case want a vacuum cleaner to hand) – take a bag of rice, pull the top of the packet wide open ... and chuck the contents straight into the air. Your aim is to eject the whole lot skyward in one jolt. Let the rice rain down.

What you have done is create a chance distribution of rice grains over the carpet. Observe the way the rice is scattered. One thing the grains have probably not done is fall evenly. There are thin patches here, thicker ones there and, every so often, a much larger and distinct pile of rice: it has clustered.

Wherever cases of cancer bunch, people demand an explanation. With rice, they would see exactly the same sort of pattern, but does it need an explanation? Imagine each grain of rice as a cancer case falling across the country. The example shows that clustering, as the result of chance alone, is to be expected. The truly weird result would be if the rice had spread itself in a smooth, regular layer. Similarly, the genuinely odd pattern of illness would be an even distribution of cases across the population.

The analogy draws no moral equivalence between cancer and rice patterns, and people who have cancer have an entirely reasonable desire to know why. But the places in which cancer occurs will sometimes be clustered simply by obeying the same rules of chance as falling rice. They will not spread themselves uniformly across the country. They will also cluster, there will be some patches where cases are relatively few and far between and others with what seem worrying concentrations. Sometimes, though rarely, the worry will point to a shared local cause. More often the explanation lies in the often complicated and myriad causes of disease, mingled with the complicated and myriad influences on where we choose to live, combined with accidents of timing, all in a collision of endless possibilities which, somehow, just like the endless collisions of those rice grains in a maze of motion, come together in one place at one time to produce a cluster.

One confusion to overcome is the belief that the rice falls merely by chance whereas illness always has a cause. This is a false distinction. 'Chance' does not mean without cause – the position of the rice has cause all right; the cause of air currents, the force of your hand, the initial position of each grain in the packet, and it might be theoretically possible (though not in practice) to calculate the causes that lead some grains to cluster. Cancer in this respect – and only in this respect – is usually no different. For all the appearance of meaning, there is normally nothing but happenstance.

Think about the rice example for a moment beforehand and there is no problem predicting the outcome. Seeing the same effect on *people* is what makes it disconcerting. This is an odd double standard: everyone knows things will, now and then, arrive in a bunch – it happens all the time – but in the event they feel put out; these happenings are inevitable, we know, yet such inevitabilities are labelled 'mysterious', the normal is called 'suspicious', and the predictable 'perverse'. Chance is an odd case where we must keep these instincts in check, and instead use past experience as a guide to what chance can do. We have seen surprising patterns before, often. We should believe our eyes and expect such patterns again, often.

People typically underestimate both the likely size of clusters and their frequency, as two quick experiments show. Shuffle a standard pack of fifty-two playing cards. Turn them from the top of the deck into two piles, red and black. How many would you expect in the longest run of one colour or the other? The typical guess is about three. In fact, at least one run of five or six, of either red or black, is likely.

You can test expectations by asking a class of schoolchildren (say thirty people) to call in turn the result of an imaginary toss of a coin, repeated thirty times. What do they think will happen? Rob Eastaway, an author of popular maths books, uses this trick in talks to schools, and says children typically feel a run of three of the same side will do before it's time to change. In fact, a run of five or more will be common. There will also be more runs of three or four than the children expect. Even though they know not to expect a uniform alternation between heads and tails on all thirty tosses, they still badly underestimate the power of chance to create something surprising.

But this is so surprising that we couldn't resist testing it: here are the real results of thirty tosses of a coin repeated three times (h = heads, t = tails). All clusters of four or more are in bold, and the longest are given in brackets.

1: h t t h t t h h t h h **t t t t t t h h h h** t h h t t t h h h
(6 tails in a row, 4 heads in a row)

2: **t t t t** h t t h h t h t **h h h h** t t t h h h t h h t h h t h
(4 tails in a row, 4 heads in a row)

3: t h **t t t t t** h t t t h h **t t t t t t** h h h t t h t h h t t
(5 tails in a row, 6 tails in a row)

There was nothing obviously dodgy about the coin either: in the first test it fell 15 heads and 15 tails, in the second 16–14, in the third 10–20, and the sequences are genuinely random.

So even in short runs there can be big clusters. The cruel conclusion applied to illness is that even if it were certain that phone masts had no effect whatsoever on health, even if they were all switched off permanently, we would still expect to find places like Wishaw with striking concentrations of cancers, purely by chance. This is hard to swallow: nine cases in twenty households – it must *mean* something. That thought encourages another: how can so much suffering be the result of chance alone? Who wouldn't bridle at an unfeeling universe acting at whim, where pure bad luck struck in such insane concentration? To submit to that fate would lay us open to intense vulnerability, and we can't have that.

But 'chance' does not mean, in the ordinary meaning of these words, spread out, or shared, or messy. It does not mean what we would think of as disordered, or without the appearance of pattern. It does not, strictly speaking, even mean unpredictable, since we know such things will happen, we just don't know where or when; and chance certainly does not mean without cause. All those cancers had causes, but they are likely to have been many and various and their concentration no more than the chance result of many overlapping and unrelated causes. It is predictable that there will be cancer clusters in a country the size of the UK, we just don't know where, and though it is a surprise when they appear, so is six consecutive tails in thirty tosses. Both ought to be expected.

Following a programme in the *More or Less* series on BBC Radio 4 about randomness and cancer clusters, in which we nervously described to one of the Wishaw villagers active in a campaign against the phone mast how clusters can occur – a woman who herself had lived with cancer – we received an email from a ferociously angry listener. How dare we, he said, take away their hope?

Chance is heartless. 'As flies to wanton boys are we to the Gods,' said Shakespeare's King Lear, 'they kill us for their sport.' More comforting amidst hurt and distress, maybe, to find a culprit you can hope to hold accountable, or even destroy.

When the surgeon and academic Atul Gawande wrote in the late 1990s about why cancer clusters in America were seldom the real thing, he quoted the opinion of the Chief of California's Division of Environmental and Occupational Disease Control that more than half the state's 5,000 districts (2,750 in fact) had a cancer rate above average. A moment's reflection tells us that this is more or less the result we should anticipate: simply put, if some are below average, others *must* be above, unless all are identical. With some grains spread out, others must be squeezed closer together. The health department in another state – Massachusetts – responded to between 3,000 and 4,000 cancer cluster alarms in one year alone, he said. Almost never substantiated, they nevertheless have to be investigated: the public anxiety is genuine and cannot be fobbed off, even if with the weary expectation of finding nothing. And these findings are not necessarily from reluctance by the medical or public health authorities to admit a problem. It is – usually – no conspiracy to suppress inconvenient facts, as the willingness to detect cancers in other ways shows. Unlike geographical clusters, discoveries of real occupational cancer clusters, or clusters of illness around a single life event – exposure to a drug or chemical for example – have been common and readily divulged. The inconvenience of the malign effects of asbestos and tobacco has not prevented the medical authorities from exposing them, despite the resistance of those industries.

Even apart from those hooked by conspiracy theories, we are all gifted recognisers of patterns and the moments when a pattern is interrupted. Something chimes or jars with an instinctive aesthetic for regularity, and we are inclined to look for causes or explanations. Whatever stands out from the crowd, or breaks from the line, invites us to wonder why. Some argue, plausibly, that we evolved to see a single cause even where there is none, on the basis that it is better to be safe than sorry, better to identify that pattern in the trees as a tiger, better to run – far better – than to assume that what we see is a chance effect of scattered light and shifting leaves in the breeze, creating an illusion of stripes. But this habit, ingrained, defiant, especially if indulged by a snap decision, makes us wretched natural statisticians. Most often the pattern *will be* a chance effect, but we will struggle to believe it. 'Where's the tiger?' we say. 'No tiger,' says the statistician, just chance, the impostor, up to her callous old mischief. In more considered moments, now we have moved on from evolution in the jungle, we should remember our experience of chance, and check the instinct born in moments in haste.

One reason for the punishing cost of medical research is that new drugs have to be trialled in ways that seek to rule out the effect of chance on whether or not they work. You might wonder how hard that can be: administer a drug to a patient and wait to see if the patient gets better, surely, is all there is to it. The joke that a bad cold treated with the latest drugs can be cured in seven days, but left to run its own course may linger for a whole week, shows us what is wrong with that. Patients improve – or not – for many reasons, at different rates. Say we put them in two groups, give one group the drug and another a placebo, and observe that the drugged group improves more. Was it the drug, or chance? Ideally we'll see a big difference between two large groups. But if the difference is small, or few people take part, it's a problem. Like rice grains or cancer cases, some patients in a medical trial will produce results that look meaningful, but are in fact quite unrelated to the influence everyone suspects – the new drug, the phone mast – and everything to do with chance.

Statistics as a discipline has made most of its progress only in the last 200 years. Perhaps the reason it took so long to get started, when science and mathematics had already achieved so much, is that it is a dry challenge to instinct. Particularly in relation to patterns, chance or coincidence, statistics can feel counter-intuitive when it frustrates a yearning for meaning.

'The tiger that isn't' makes a good standard for numbers that seem to say something important but might be random, and there are imaginary tigers in all walks of life. Our question could be this: is the tiger real? Or are we merely seeing stripes? Is this a pattern of numbers that tells us something, or a purely chance effect that bears only unsettling similarity to something real?

Like illness, events cluster too. In 2005 three passenger airliners came down in the space of a few weeks, leading to speculation of some systemic problem – 'What is causing our planes to fall?' To repeat, chance does not mean without cause – there was a cause for each of those crashes – just separate causes. What chance can do is to explain why the causes came together at the same time, why, in effect, they clustered.

Does this prove that every cluster, cancer or otherwise, is chance alone? Of course not, but we have to rule out that explanation before fastening on another. People in suits seen on the news advising disbelieving residents that their fears are unfounded might be part of a conspiracy against the public interest – it is conceivable – but let's also allow that they speak from acquaintance with what chance is capable of, and have worked honestly to tell stripes from the tiger. The difference often comes down to nothing more than size. One case of a real cluster, in High Wycombe, of a rare type of nasal cancer with a genuinely local cause, was eventually attributed to the inhalation of sawdust in the furniture industry, and had prevalence 500 times more than expected.

But clustering is only part of chance's repertoire of confusion. She can also cause numbers to rise and fall. Seeing them go up or down, people fidget for meaning, as they do for clusters – what made them rise; what made them fall?

But numbers can go up and down for no more reason than occurs when chance varies the height of the waves rolling into shore. Sometimes waves crash, sometimes they ripple, now larger, now smaller, varied on days even when the wind blows steadily.

This succession of irregular peaks and troughs, swelling and sinking, fussy and dramatic, we'll call chance. Like chance it has its causes, but so intricately tangled we can almost never unravel them. Imagine those waves like numbers that are sometimes high and sometimes low. It is the easiest mistake in the world to observe a big wave, or a high number, and assume that things are on the up. But we all know that a big wave can arrive on a falling tide. This is no great insight. Everyone knows it from life. The suggestion here is simply that what is known in life is used in numbers. The tendency among some to feel that numbers are alien to experience, incomprehensible, inaccessible, a language they will never speak, should be put aside. In truth, they speak it already. Numbers often work on principles that we all use with easy familiarity, all the time. And so we can easily see that a momentarily high number – in accident rates or elsewhere – can, just like a big wave, be part of a falling trend. And yet taking a measurement from a single peak and assuming it marks a rising tide, though a simple error, is one that in politics really happens. It is assumed that the peak in the numbers tells us all we need to know, that it happened for a critical reason, caused by an action we took or a policy we implemented, when really it is just chance that made the numbers momentarily swell. Is this obvious? It is hard to believe that politics sometimes needs lessons from the rest of us in the view from the beach, but it is true. For to know if the explanation is accurate, what matters is not the size of the wave at all, but rather the tide, or maybe even sea level. The wave catches the eye, but look elsewhere before being sure of capturing a real or sustained change, not a chance one. Otherwise, you will fail to distinguish cause and effect from noise.

The inevitability of such chance variation in numbers like this regularly sinks understanding. A contentious and none-too-obvious example is the roadside speed camera. These are deadly devices, if not to traffic then certainly to conversation. Oppose them and you crave toddlers at play slaughtered; you seek to scream unimpeded through residential areas pushing 90mph. Support them, and it is owing to your control-freak loathing for others' freedom and pig-ignorance of the facts about a flawed system. The antis are caricatured as don't-give-a-damn evaders of civic responsibility, the pros of wanting the motorist fined off the road by a stealth tax to fill the coppers' coffers. According to the *Daily Telegraph* (10 August 2006), 'a survey found that 16 per cent of people support the illegal destruction of speed cameras by vigilante gangs'.

By Luke Traynor – *Liverpool Daily Post*
MERSEYSIDE'S speed cameras are picking up motorists driving at up to 134mph, figures revealed yesterday. In one incident, a driver was caught racing at 134mph in a 50mph zone, nearly three times over the limit, on the M62 near the Rocket public house. Over a six-month period 116 motorists were caught going at or above 70mph in a 30mph zone.

Three cheers for speed cameras? Then again …

By Philip Cardy – *Sun*
Nicked for doing 406mph.
Driver Peter O'Flynn was stunned to receive a speeding notice claiming a roadside camera had zapped him – at an astonishing 406mph. The sales manager, who was driving a Peugeot 406 at the time, said: 'I rarely speed and it's safe to say I'll contest this.'

With indignant voices on either side, sane discussion is perilous, but why is there even argument? Judgement ought easily to be had from the numbers: are there more accidents with cameras or without? The answer seems clear: on the whole, accidents decline when cameras are installed, sometimes dramatically.

Department for Transport Press Release, 11 February 2003
Deaths and serious injuries fell by 35% on roads where speed cameras have been in operation, Transport Secretary Alistair Darling announced today. The findings come from an independent report of the two-year pilot scheme where eight areas were allowed to re-invest some of the money from speeding fines into the installation of more cameras and increased camera use.

> Transport Secretary Alistair Darling said: 'The report clearly shows speed cameras are working. Speeds are down and so are deaths and injuries ... This means that more lives can be saved and more injuries avoided. It is quite clear that speeding is dangerous and causes too much suffering. I hope this reinforces the message that speed cameras are there to stop people speeding and make the roads safer. If you don't speed, you won't get a ticket.'

This 35 per cent cut in deaths and serious injuries equated, the department said, to 280 people, and since then the number of cameras has grown. Case closed? Not quite. First, there are exceptions: at some sites the number of accidents went up after speed cameras arrived. But a minority of perverse results will not win the argument, we need to know what happens on balance: on balance there was still a large fall.

The second reservation risks sounding complicated, but is in principle easy: it is the need to distinguish between waves and tides (which for some years the Department for Transport preferred not to do). As with waves, the rate at which things happen in the world goes up and down, though without telling us much about the underlying trends: there are more this time, fewer next, rising one week, falling another, simply by chance. People especially do not behave with perfect regularity. They do not sort themselves into an even flow, do not get up, go to work, eat, go out, catch the bus, crash, either at the same time or at even intervals. The numbers doing any of these things at any one time goes up and down.

Of course, you knew that already. Accident statistics at any one site will quite likely go up and down from time to time, because that is what they do – a bad crash one month, nothing the next. The freakish result would be if there were exactly the same numbers of accidents in any one place over every twelve-month period. Luck, or bad luck, often determines the statistics, as the wave rises and falls.

The argument bites with the realisation that, all else being equal, if the numbers have been higher than usual lately, the next move is more likely to be down than up. After a large wave often comes a smaller one. Two bad crashes last month on an unexceptional stretch of road and, unless there is some obvious hitch, we would be surprised if things did not settle down.

Applied to road accidents, the principle has this effect: put a speed camera on a site where the figures have just gone up, at the crest of a big wave – and this does tend to be when the cameras arrive, in the belief that these sites have been newly identified as a problem – and the next move in the accident statistics is likely to be down – *whether the speed camera is there or not*. The peak turns to a trough, the statistics fall and presto, the government claims success.

A small experiment shows how pure chance can produce what looks like evidence in favour of speed cameras, using a simple roll of a die. Before describing it, we had better state our view lest we are labelled extremists of one sort or another. On the statistical evidence, most speed cameras probably do cut accidents, some probably do not, depending where they are. So there is probably a benefit, but the size of that benefit has been greatly exaggerated by some – including the government. And this takes no account of the effect that withdrawing police patrols in favour of using roadside cameras has on accidents caused by other driving offences – drink-driving for example – an effect extremely difficult to estimate. (See also Chapter 7 on risk.)

So, to the experiment. Find a group of people – a typical

class size will do but it can be smaller – and ask each person to be a stretch of road, as we did with a gaggle of journalist colleagues. Take your pick: the A42 past Ashby De La Zouche, the B1764 to Doncaster, all perfectly ordinary stretches of road. Next, everyone rolls a die twice and adds the two throws together. This represents the number of accidents on each stretch of road. And, just by chance, some stretches produce higher numbers of accidents than others, as they do in life when brakes or a driver's attention, for example, just happen, by chance, to fail at that moment. In our group of about twenty volunteer journalists we then targeted what appeared to be the accident black-spots, by identifying all those with a score of 10, 11 or 12. To these we gave a photograph of a speed camera, and asked them to roll again, twice. The result? The speed camera photos are instantly effective. None of the high scorers equalled their previous score once they had a photograph of a speed camera.

It might be objected that these were not the real thing, only photographs, and so the experiment proves nothing. But the point is that at some genuine roadside accident sites placing a photograph – or even pebble – on the pavement would have been as effective as a camera, simply because the rise and fall in the numbers of accidents has been due to chance, just like a six on the roll of a die. Since we have put the cameras (or the photographs of a camera, or the pebble) in a place where the number has just, by chance, been high, it is quite likely, whatever we do, by chance to go down. It looks as if we had something to do with it. In fact, we just caught the wave and took the credit.

This is a phenomenon known to statisticians as 'regression to the mean' – when a number has reached a peak or trough lately, the next move is likely to be back towards the average

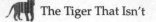

Figure 2 **Accidental success**

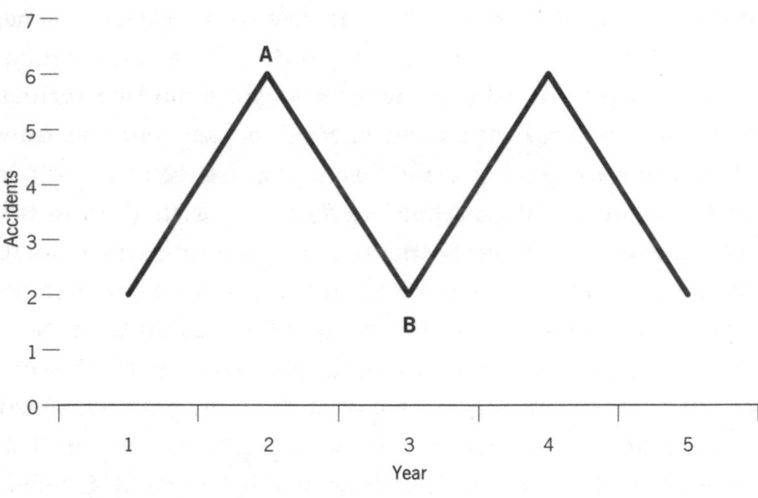

– a phenomenon initially discounted by Department for Transport researchers, despite our challenge to their figures. In the chart, the accident rate goes up and down, as with waves, as in life. Imagine cameras are put in place at time A, when accidents just happened to be high, then it's noticed that they fall to B. This was going to happen anyway, but because the cameras are there it is all too easy to claim that the cameras did it. That is effectively what the Department for Transport did. Then, after claiming for two years a large number of lives saved by cameras, estimates of the benefit were cut sharply in 2006 when, for the first time, some attempt was made to calculate how much of this was due to regression to the mean, and chance was finally given her due.

Even after revision the DfT figures were still unsatisfactory, being the result of a dubious mixture of definitions and some sloppy arithmetic. Figures in the ministerial press release had to be corrected after we challenged them. But in their more modest claims they were at least probably a little closer to the truth.

Of the full fall in the number of people in fatal or serious collisions at speed camera sites, about 60 per cent was now attributed to regression to the mean, and about another 18 per cent to what are called 'trend' effects, namely that the number of accidents was falling everywhere, including where there were no cameras, as a result, for example, of improved road layout and car safety. This leaves, according to the department's revised figures, about 20 per cent of the apparent benefit of speed cameras genuinely attributable to them, though even this figure is contested.

To observe the tide and not the wave of accidents at speed-camera sites, there are two options: (a) wait until we have seen plenty of waves come and go, which in the case of speed cameras is now believed to be for about five years; (b) do not begin measurements only from the peaks, but do plenty of them, and choose points of measurement at random. More simply, we can remember that there is a lot of chance about.

Public policy worldwide has a truly shocking history of ignorance about whether the benefits it claims in a precious few oft-quoted examples have occurred entirely as a result of chance or other fortuitous association, or if the policy genuinely makes the differences claimed. At the Home Office, for example, a report in early 2006 on the evidence for the effectiveness of policy to tackle reoffending found that not one policy reached the desired standard of proof of efficacy, because so many had failed to rule out the possible effects of chance when counting the rise or fall in offences by people on various schemes. A senior adviser at the Home Office who knows what a sneaky devil chance can be, and how easily the numbers can mislead, says to ministers who ask what actually works to prevent reoffending: 'I've no idea.'

Can this really be how government proceeds? Without much by way of statistical rigour and cursed with a blind spot for the effects of luck and chance? All too often, it is.

In the United Kingdom, that tendency to ignore the need for statistical verification is only now beginning to change, with the slow and often grudging acceptance that we need more than a plausible anecdote (a single wave) before instituting a new policy for reoffenders, for teaching methods, for healthcare, or any other state function. Politicians are among the most recalcitrant, sometimes pleading that the genuine pressure of time, expense and public expectation makes impossible the ideally random-controlled trials which would be able to identify real stripes from fake, sometimes apparently not much caring or understanding, but, one way or another, often resting their policies on little more than luck and a good story, becoming as a result the willing or unwilling suckers of chance.

A politician with a taste for a calculated gamble is a disappointingly welcoming way in for chance to do its dirty work. A more shocking example comes with the two innocent doctors who felt confronted – on nothing more than the turn of a roulette wheel – with an insinuation of murder.

It has happened. Two general practitioners, summoned to a meeting, sat tense and anxious, trying to explain the unusual death rates of their patients. The meeting took place in the shadow of Dr Harold Shipman – the recently convicted Yorkshire GP who murdered probably well in excess of 200 people. An inquiry into the Shipman case, led by Dame Janet Smith, was trying to establish whether GP patient mortality could be monitored in order to spot any other GPs who gave cause for concern. These two doctors had been identified from a sample of about 1,000 as among a dozen whose patients were dying at rates as high as, or higher than Shipman's.

An initial statistical analysis had failed to find an innocent explanation. What other remained, except for quality of care – for which read the hand of the two GPs? This was where Dr Mohammed Mohammed came in. He was one of those asked to investigate. 'This was not a trivial meeting', he told us. One can imagine. It was deeply uncomfortable, but necessary, he added, in order to explore any plausible alternatives. He was not about to throw at anyone a casual accusation of murder, but they had all read the newspapers. A softly spoken, conscientious man, determined to proceed by scientific method, he wanted to know before coming to a conclusion if there were any other testable hypotheses.

The statistical analysis that brought the GPs to this point had sought to sift from the data all variation due to chance or any unusual characteristics of their patients: the ordinary rise and fall in the numbers dying that would be expected anywhere, together perhaps with an unusually elderly population in the area, or some other characteristic in the case-mix that made deaths more frequent, were obvious explanations. But these were examined and found wanting. Now it was the GPs' turn.

And the reason they offered, essentially, was chance; chance missed even by a statistical attempt to spot and control for it. No, their patients were not significantly older than others, but the GPs did have, by chance, an unusually high number of nursing homes on their patch. And while age alone is often a good proxy for frailty or increased illness, a nursing home is much better. People here are most likely to be among the most frail or in poor health.

We know without doubt that there was nothing accidental about Shipman, but the variation of death rates, even variation on a par with mass murder, can have a perfectly innocent explanation, as chance ruffles the smoothness of events, people and places, changing a little here and there, bringing a million and one factors into play, suggesting vivid stripes without a tiger to be seen.

To see the problem of telling one from another, think of your signature. It varies, but not too much, and what variation there is worries no one, it is normal and innocent. Try with the other hand, though, and the degree of variation probably jumps to the point where you hope that on a cheque the bank would reject it. The task in detecting a suspect pattern is to know where the normal chance variation stops, and what is known as 'special cause' variation begins. How varied can a signature be and still be legitimate? How much variation in mortality rates can there be before we assume murder? The answer is that innocent or chance variation can exceed even the deliberate effect of Britain's most prolific serial killer.

The GPs' explanation turned out to be entirely consistent with the data, once examined closely, mortality rates at the nursing homes being perfectly in line with what would be expected – for nursing homes. They were cleared completely, but hardly happy. If the sample of GPs that identified these two was anything to go by, we would expect to find about 500 across the UK with mortality rates similarly inviting investigation. That is an awful lot of chance at work, but then, chance is like that: busy – and cunning, and with about 40,000 GPs to choose from in the UK, some would certainly be unlucky even if none was murderous.

Dame Janet Smith said in her final report from the Shipman inquiry: 'I recognise, of course, that a system of routine monitoring of mortality rates would not, on its own, provide any guarantee that patients would be protected against a homicidal doctor. I also agree with those who have emphasised the importance, if a system of routine monitoring is to be introduced, of ensuring that PCTs (Primary Care Trusts), the medical profession and the public are not lulled into a false sense of security, whereby they believe the system of monitoring will afford adequate protection for patients.'

Aware of these difficulties, she recommended GP monitoring all the same, but justified as much by the value of the insight it might offer into what works for patient care – a view shared by Dr Mohammed – as for detecting or deterring murder. Can we really not tell chance death from deliberate killing? With immense care, and where the numbers are fairly clear-cut, as in the Shipman case, we can, just about. But he was able to avoid detection for so long because the difference was not instantly obvious then and, with all our painfully learned statistical wisdom since, the chilling conclusion is that we are only a little better at spotting it now, because chance will always fog the picture. It was in February 2000, shortly after Shipman's conviction, that the then Secretary of State for Health, Alan Milburn, announced to Parliament that the department would work with the Office of National Statistics 'to find new and better ways of monitoring deaths of GPs' patients'. This promise was still unfulfilled seven years later. That is a measure of the problem – and testimony to the power of chance.

When statisticians express caution, this is often why: chance is a dogged adversary. And yet it can be overcome. Trials for a new polio vaccine in America and Canada in the 1950s had to meet persistence with persistence. At that time, among 1,000 people, the chances were that none would have polio. It always was quite rare. So let's say a vaccine comes along and it is given to 1,000 people. How do we know if it worked? Chance would probably have spared them anyway. How can we tell if it was the vaccine or chance that did it?

The answer is that you need an awful lot of people. In the trials for the Salk polio vaccine, nearly 2 million children were observed in two types of study. The number who for one reason or another did not receive the vaccine, either because they were in one of the various control groups, or simply refused, was 1,388,785. Of these, 609 were subsequently diagnosed with polio, a rate of about one case in every 2,280 children.

Among the nearly half million who were vaccinated, the rate was about one case in nearly 6,000 children. The difference was big enough, among so many people, that the research teams could be confident they had outwitted chance's ability to imitate a real effect. Though even here they looked closely to ensure there were no chance differences between the groups that might have accounted for different rates of infection. Getting the better of chance can be done, with care and determination, and often involves simply having more stamina or patience than she does.

One more example, which we will look at in more detail in a later chapter. School exam results go up and down from year to year. They move so much, in fact, that the league table is shuffled substantially each year. But is it the school's teaching standards that are thrashing up and down? Or is the difference due to the ups and downs in pupils' ability as measured by exams from one year to the next? It seems mostly the latter, and that is as you might expect: what principally seems to determine a school's exam results is the nature of its intake. In fact, the results for a school in any one year are so subject to the luck of the intake that for between two thirds and three quarters of all schools, the noise of chance is a roar, and we are unable to hear the whisper above it of a real influence or a special cause; we are unable to say with any confidence whether there is any difference whatsoever that is made by the performance of the schools themselves. Chance so complicates the measurement that for the majority of schools the measurement is, some say, worthless. We would not go nearly that far. But we would agree that data is regularly published that misleadingly implies these waves are a reflection of the educational quality of the school itself, and report year-to-year changes in performance as if they were clear indicators of progress.

A rising tide or just a wave? Stripes or a real tiger? We can be vigilant, we need to be vigilant, but we will be fooled again. The least we can do is determine not to make it easy for chance to outwit us. That task is begun by knowing what chance is capable of, a task made easier if we slow the instinctive rush to judgement and beware the tiger that isn't.

5

Averages: The White Rainbow

Averages summarise; that is their essence. But in saying a little about a group, they can obscure all that matters about its parts. They are suggestive, but sketchy, like a bland, one-phrase round-up of the evening news, a mere lick and a promise from a world of infinite differences. The abiding problem with the average is that it stifles imagination about the hotchpotch world it claims to speak for. The way to see through it is to remember that world's variety.

The average wage, average house prices, average life expectancy, average crime rate, as well as less obvious averages like the rate of inflation; these and others reduce all to one, summarising a mass of data in a single number, one point on the graph, one value supposedly speaking for all, when what they represent may well be a sprawling mess, encompassing the whole glory and the shame of human experience, the blessing and misery, the typical and the odd, wantonly distributed, some here, some there, some almost off the graph, but all somehow miraculously blended into unity by this thing called the average. For these reasons, it is both potent and flawed: it tidies up and smooths over; flattens hills and raises hollows to tell you the level of the land – were it flat.

But it is not flat. Forget the variety behind every average and you risk trouble, like the man who drowned in a river that rose, he heard, on average only to his knees. There was a variety of depths he thought he had the measure of. Always try to picture that variety.

Two images might help: the sludgy black/brown that children make when petulant with the paint pots is a kind of average – of the colours of nature – and no one needs telling it is a deceptive summary of the view. 'White, on average,' is what we'd see by combining the light from a rainbow, then sharing it equally. But this bleeds from the original all that matters – the magical assortment of colours. Whenever you see an average, think: 'white rainbow', and imagine the colours it conceals.

Victoria Lacey gave birth to baby Sasha on 10 September 2005. Her due date had been 26 August. Two weeks late and pregnancy now maddening, she began each day with hope this would be the one and then, as the long hours passed, resigned herself to another. Was something wrong? 'Why can't your body produce a baby on the date it's supposed to?' she asked herself.

But which date is that? Doctors give expectant mothers an estimated date since, naturally, they can never be certain, and that estimate is based on the average length of a pregnancy. But how long is the average pregnancy? The answer, unhelpfully, is that the official average pregnancy is shorter than it probably ought to be.

There were 645,835 live births recorded in the UK in 2005; is it possible that every due date was misleading? Of course, some will have been right by chance simply because pregnancies vary in duration, but will they have been right as often as they could have been? The impression of an imprecise science is confirmed when we learn that the practice in France is to give a latest date, not a due date, some ten days later than in Britain.

Due dates in the UK are initially calculated by counting 280 days from the first day of the last menstrual period. British doctors settled on this number in part because it seemed about right, but also under the influence of a Dutch professor of medicine named Gustaave Boerhaave ('So loudly celebrated, and so universally lamented through the whole learned world' – Samuel Johnson).

Boerhaave wrote nearly 300 years ago that the duration of pregnancy was well known from a number of studies. Those studies have not survived, though their conclusion has. It remains well known up to the present day, consolidated by influential teachers and achieving consistency in medical textbooks by about the middle of the twentieth century. Nearly everyone in the UK still agrees that 280 days is correct: it is the average.

But averages can deceive. A drunk sways down the street like a pendulum from one pavement to the other, positioned on average in the centre of the road between the two white lines as the traffic whistles safely past, just. On average, he stays alive. In fact, he walks under a bus.

Averages put variation out of mind, but somewhere in the average depth of the river there might be a fatally deep variation; somewhere in the distribution of the drunk's positions on the road there was a point of collision that the average obscures. The enormous value of the average in contracting an unwieldy bulk of information to make it manageable is the very reason why it can be so misleading.

To use another metaphor, the world is a soup of sometimes wildly varying ingredients. The average is like the taste and tells us how the ingredients combine. That is important, but never forget that some ingredients have more flavour than others and that these may disguise what else went into the pot. If we are to avoid coming away with the idea that the average English vegetable tastes of garlic, we also need to know the whole recipe. Averages stir traces of the richer, weirder world into one vast pot with everything else, mix the flavours and turn the whole into something which may or may not be true of most people, and may be true of none.

If you find that thought hard to apply to numbers, simply remember that any average may contain strong flavours too: distorting numbers, atypical numbers. Think, for example, of the fact that nearly everyone has more than the average number of feet (yes, nearly everyone can have more than the average). This is so because a few people have just one foot, or no feet at all, and so the tiny influence of a tiny minority is nevertheless powerful enough to shift the whole average to something a bit less than two. To have the full complement of two feet is therefore to be above average. Never neglect what goes into an average, and why it might move the average somewhere misleading.

So we need to go carefully when we are offered an average as a summary of disparate things. It would help if journalists, politicians and others took care to avoid using them as a way of saying 'just about in the middle' (unless this is an average called the median), avoided using them to stand for what is 'ordinary', 'normal' or 'reasonable', avoided using them even to mean 'most people', unless they are sure that is what they meant. They may be none of those things.

So what is going on in the rich variety of experiences of pregnancy? In particular, what happens at the edges? Two facts about pregnancy suggest that the simple average will be misleading. First, some mothers give birth prematurely.

Second, almost no one is allowed to go more than two weeks beyond the due date before being induced. Premature births pull the average down; late ones would push it up, but we physically intervene to stop babies being more than two weeks late. The effect of this imbalance – we count very early births but prevent the very late ones – is to produce a lower average than if nature were left to its own devices. That is not a plea to allow pregnancies to continue indefinitely, just to offer a glimpse inside the calculation.

We might argue in any case that very premature births ought not to be any part of the calculation of what is most likely. Most births are not significantly premature, so if a doctor said to a woman: 'The date I give you has been nudged down a few days to take account of the fact that some births – though probably not yours – will be premature,' she might rightly answer, 'I don't want my date adjusted for something that probably won't happen to me, I want the most likely date'. The average is created in part by current medical practice – the numbers we've got arise partly through medical intervention – and the argument in favour of 280 days becomes circular: this is what we do because it is based on the average, and this becomes the average partly because of what we do.

Mixing in the results from the edges produces a duration that is less likely to be accurate than it could be. In the largest recent study, of more than 400,000 women in Sweden, most had not yet given birth by 280 days – the majority of pregnancies lasted longer. By about 282 days, half of babies had been born (the median), but the single most common delivery date (the mode), and thus arguably the most likely for any individual in this distribution, was 283 days.

If most women have not had their baby until they are at least two days overdue, and women are more likely to be three days overdue than anything else, it invites an obvious question: are they really overdue?

You are forgiven for finding it muddled. Yet above all this stands the unarguable fact, confirmed by all recent studies, that the duration at which more women have their baby than any other is 283 days. This type of average – the most common or popular result – is known as the mode, and it's not clear why in this case it isn't preferred. In fact, that latest study from Sweden found that even the simple arithmetical average (the mean) was not in fact 280 days, but 281.

None of this would matter much (one, two or three days is sometimes frustrating, but in a normal pregnancy most likely to be medically neither here nor there at this stage), were it not that these numbers form part of the calculation of when to induce birth artificially. Induction is often offered to women in this country, sometimes with encouragement, seven days after the due date. It raises significantly the likelihood of Caesarean section – which has risks of its own – and can be, for some, a deeply disappointing end to pregnancy.

When obstetricians answer that induced birth has better outcomes for women than leaving them a little longer, they fail to tell you that one of the ways they measure that 'better outcome' is by asking women if they feel better, having sorted out the problem of being overdue. If you tell someone there's a problem, they may well thank you for solving it. If they knew the problem was based on a miscalculation, they might feel otherwise.

Averages bundle everything together. That is what makes them useful, and sometimes deceptive. Knowing this, it is simple to avoid the worst pitfalls. All you need do is remember to ask: 'What interesting flavours might be lost in that average? What else could be in the pot? And if they tell me the rainbow is white on average, my imagination for what is blanked out must be my guard.'

Averages are sometimes credited with authority they don't deserve, largely because they are thought to equate (or used as if they do) with 'normal', 'typical', or 'reasonable'. In the 2005 general election, Charles Kennedy, then leader of the Liberal Democrats, found himself ducking the notion of the average income like a sword of truth, swung in defence of the ordinary citizen.

His party had said we should be rid of council tax, replacing it with a local income tax. Some – the richest – would pay more, most would pay less. Then came the inevitable question: how much did you have to earn before local income tax raised your bill? As Charles Kennedy stumbled, at a loss for the accurate figure, the journalists at the press conference sniffed their story. The answer, when it came in a scribbled note from an adviser, gave the Liberal Democrats their roughest ride of the campaign: A little over £40,000 per household.

The problem was that this figure turned out to be about twice average earnings. To take the example every news outlet seized, a firefighter on average earnings living with a teacher on average earnings would pay more under the Liberal Democrats' proposed new system.

The campaign trail was suddenly hazardous for Charles Kennedy, the questions and coverage indignant: what was fair about a system that hit people on average earnings? If it is going to be more expensive for the average, how can you possibly say most people will be better off? So, the Lib Dems are hitting Middle England? And so on.

There might be other good reasons for opposing the policy, but there is also such a ragbag of misconceptions about the idea that lay at the root of the criticism – the average earner – that it is hard to know where to start.

As with pregnancy, so with income: the average is not in the middle, and more people are on one side than the other. That's more straightforward than it sounds. Think again of how the average number of feet is very slightly less than two (it only takes one person in the world with one foot to bring the average down – if we assume no one has three), so anyone with two feet – that is almost everyone – has more than average. One strange ingredient changes the average of the whole.

With income most people earn less than the average. The relatively few above the average have, metaphorically speaking, more feet than the rest, and their effect is to pull the average well above the middle. With earnings, that effect is magnified. There are quite a few people who are income centipedes. Put them into the mix with everyone else and they're powerful enough to shift the average so far that the majority of earners find themselves below average.

Because the average is pulled higher in this way, leaving most people below it, a household with two individuals each earning average individual earnings is not in the middle of the household income distribution, it is, in fact, in the top quarter; it is, if you are wise to the rainbow distribution of incomes and not cocooned by the even larger salary of a national newspaper commentator, to be relatively rich. This is for two reasons: first, because in this case the average is pulled well beyond the middle by a relatively small number of enormously high incomes; second, because it is relatively unusual for both members of a couple to have average earnings. In most cases where one half of a couple has average earnings, the other will earn significantly less. It might surprise us to find that a teacher and a firefighter living

Figure 3 **Who earns what?**

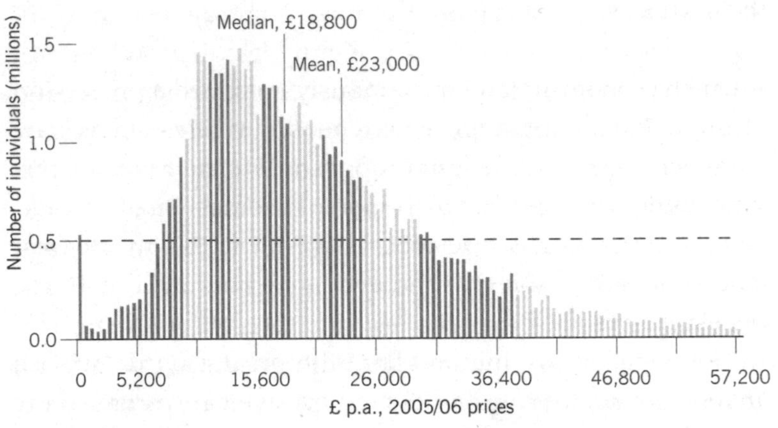

together are relatively rich, but only if we do not know how they compare with everyone else, don't know where they are in the distribution, and have ignored the colours of the income rainbow.

Figure 3 shows the distribution of income in the UK for childless couples – two people living together, their incomes combined. Half have net incomes (after tax and benefits) of less than £18,800 (marked as the median), but the average for the group is around £23,000, pulled up by the relatively small numbers of very high incomes. That is, most are more than 18 per cent below average. The highest incomes are far too high to fit on the chart, which would need to stretch many feet to the right of the edge of the page to accommodate them. The most common income is around £14,000, about 40 per cent below average. More people in this category have incomes at this level than any other. If that is a shock, it is probably due to the familiarity of the much higher and more widely quoted average. Most media or political comment about 'middle Britain' does not have a clue, economically speaking, where the middle is.

A Dutch economist, Jan Pen, famously imagined a procession of the world's population where people were as tall as they were rich, everyone's height proportional to their wealth, this time, rather than income. A person of average wealth would be of average height. The procession starts with the poorest (and shortest) person first and ends, one hour later, with the richest (and tallest).

Not until twenty minutes into the procession do we see anyone. So far, they've had either negative net worth (owing more than they own) or no wealth at all, and so have no height. It's a full thirty minutes before we begin to see dwarfs about six inches tall.

And the dwarfs keep coming. It is not until forty-eight minutes have passed that we see the first person of average height and average wealth, when more than three quarters of the world's population has already gone by.

So what delays the average so long after the majority have passed? The answer lies in the effect of the people who come next. 'In the last few minutes,' wrote Pen, 'giants loom up … a lawyer, not exceptionally successful, eighteen feet tall'. As the hour approaches, the very last people in the procession are so tall we can't see their heads. Last of all, said Pen (at a time before the fully formed fortunes of Bill Gates and Warren Buffett), we see John Paul Getty. His height is breathtaking, perhaps ten miles, perhaps twice as much.

One millionaire can shift the average more than many

hundreds of poor people, one billionaire a thousand times more. They have this effect to the extent that 80 per cent of the world's population has less than average.

In everyday speech, 'average' is a word meaning low or disdained. With incomes, however, the average is relatively high. The colloquial use, being blunt, thoughtless and bordering on a term of abuse, distorts the statistical one, which might, according to the distribution, be high, or low, or in the middle, or altogether irrelevant. We must have a little background before we know which.

The writer and palaeontologist Stephen Jay Gould had special reason to find out. Diagnosed with abdominal mesothelioma in 1982, a rare and highly dangerous cancer, at the height of a brilliant career and with two young children, he quickly learned that it was incurable and that the median survival time after discovery was eight months. He gulped, he wrote later, and sat stunned for fifteen minutes.

Then, he says, he started thinking. He describes the ensuing story – a statistical one – as profoundly nurturing and life-giving.

The median is another variety of average: it means the point in a distribution where half are above and half below, what Gould and others call a central tendency. In this case it meant that half of all people diagnosed with abdominal mesothelioma were dead within eight months.

A number like that hits brutally hard. But it is worth recalling that other numerical characteristic which applies particularly to averages of all kinds, namely, that they are not hard and precise, do not capture some Platonic essence, are not, as Stephen Jay Gould understood to his advantage, immutable entities. Rather, averages are the classic abstraction, and the true reality is in the distribution, or as Gould says: '... in our actual world of shadings, variation and continua'.

In short, wrote Gould, 'we view means and medians as the hard "realities" and the variation that permits their calculation as a set of transient and imperfect measurements'. In truth, so far as there are any hard realities at all, they are in the variation, in the vibrant individual colours of the rainbow, not in the abstracted average that would declare the rainbow white.

Once again, to make sense of this average, we must ask what is in the distribution. The first part of the answer is straightforward: half will live no more than eight months, that much is true, but half will also live longer, and Gould, being diagnosed early, reckoned his chances of being in the latter half were good.

The second part of the answer is that the eight-month median tells us nothing about the maximum for the half who last longer than eight months. Is eight months half-way, even for the luckiest? Or is the upper limit, unlike the lower one, unconstrained? In fact, the distribution had a tail that stretched out to the right of the graph for several years beyond the median, being what statisticians call right-skewed. If you lived longer than eight months, there was no knowing how much longer you might live beyond that.

Understanding the distribution behind the average allowed Gould to breathe a long sigh of relief: 'I didn't have to stop and immediately follow Isaiah's injunction to set thine house in order for thou shalt die, and not live,' he wrote. 'I would have time to think, to plan, and to fight.' He lived, not eight months, but another 20 years, and when he did die, in 2002, it was of an unrelated cancer.

The same skewed distribution lies behind the average

life span for people in Swaziland, the lowest in the world according to the 2007 CIA Fact-book and United Nations social indicators, at 32 years for men and 33 for women. It is a frighteningly low figure. But 32 is not a common age of death – most who make it that far survive longer – it is simply that in order to calculate the average their fates are bundled together with the shocking number who die in infancy. The average is pulled down by Swaziland's high rates of infant mortality. And so the figure for average life expectancy stands like a signpost between two destinations, pointing at neither. It fails to convey either of what we might call the two most typical life spans, but is an unsatisfactory blend of something truly awful and something more hopeful, a statistical compromise that describes very few. A better way of understanding life expectancy in Swaziland would be to say that it is either only a few years, or that it is not very far from our own. Prospects there are largely polarised; the average yokes the poles together.

Should averages simply be avoided? Probably not, because for all their hazards, sometimes we want a figure to speak for a group as a whole. And averages can be revelatory. Making them useful is mainly a question of working out which group you are interested in. What we most often want to know when we find ourselves reaching for an average about some social, economic or political fact of life, is what is true for most people or most often, what is typical, what, to some extent, is normal for a particular group. Some of the many reasons an average might not tell us any of these things are found above, to do with the awkward failure of life to behave with regularity. But say we are pig-headed and still want to know. Then there is nothing for it: with all those caveats about what might be going on in the misty reaches of the distribution, we still need to find words for a summary. And sometimes that summary of the typical will be what we most want to know about the success or failure of a policy.

A good current example is hospital waiting lists. The government has set a target – currently that no one should wait more than six months (twenty-six weeks) for an operation. This is about to change to the much more demanding target of a maximum of thirteen weeks from GP referral to treatment, continuing the deliberate focus on one end of the distribution, the long-waiting end. Once there was a long tail to the waiting list, and you could wait years for treatment, but it is now an abrupt cut-off with a maximum waiting time for all. It reminds us how certain parts of the distribution rather than the whole can acquire political importance.

As a result of this policy, the government has been fond of saying that waiting times are coming down (Department of Health press release, Wednesday 7 June 2006: 'Waiting times for operations ... shorter than ever'), and it is quite true that the longest waits have been reduced dramatically. But the longest waits, though a real problem, are a small proportion of the total number of waits. Not very many were ever kept waiting more than two years compared with the millions who were seen within a few months. So this is a case where we might also want to know, in order to say with authority what is happening generally to waiting times, what is happening to everyone else, not only to the long waits, but also to the more typical waits, those who always were inside the maximum. The government has taken one part of the distribution and talked about it as if it spoke for the whole. How do we find a more satisfactory measure?

The best answer in this case is to ask what happens to a patient whose wait is somewhere in the middle of all waits, the point where half of all patients wait less and half more, and which is an average called, as in the Stephen Gould case, the median. These figures are decidedly mixed, and in some cases startling. Even so, this example will not take the median for every patient in the country, but will identify the median for various large groups.

Before we look at them, we will also do one other thing to the calculation of waiting times. We will strip out a practice that seems to have increased significantly in recent years of resetting the waiting-time clock for what are sometimes opportunistic reasons (see Chapters 3 and 6 on counting and targets). This is permitted but easily abused – the hospital rings you to offer an appointment with one day's notice, you can't go, they reset your waiting time. Once you strip out this behaviour to compare like with like waits, five years ago with now, and look at the typical patient, not just the very longest waits, the effects are illuminating.

In one primary care trust (PCT) area we looked at (in late 2006), for example, waiting times for trauma and orthopaedics had gone up for the typical patient from 42 days to 102 days. In another there was an increase from 57 days to 141 days. A third saw a rise from 63 days to 127 days.

Waiting times for ear, nose and throat (ENT) told a similar story, where the typical patient in about 60 per cent of primary care trusts is waiting as long or longer than five years ago. In general surgery, figures for more than half of all PCTs showed that the typical patient was waiting longer than five years ago. In a number of cases, though not a majority in the areas we looked at, even the waits for the 75th percentile (how long before 75 per cent of patients are treated) had gone up.

The key point of our analysis is that it makes little sense to say waiting times are going one way or another unless you say 'for whom?' and identify the group that interests you. The figures are not moving the same way for everyone, and in some large categories of treatment, though not for all, most patients, as typified by the median, are waiting as long or longer than five years ago.

It is important to know what has happened to the longest waits and it is quite reasonable to say that shortening them is a success for those patients, but it makes no sense to say that these alone tell you whether 'waiting times have come down'. They do not. For this, there is no option but to use some kind of average, and the most appropriate average is the median.

Incidentally, we found that most hospitals we asked could not tell us what was happening to the typical (median) patient, and the data came from another source – the Dr Foster organisation – which had access to the raw hospital episode statistics and the in-house capacity to crunch the data.

Always ask about an average: which group are we really interested in? Maybe when we ask about the average income, we don't want to know about the stratospheric earners, we want to know about what's more typical. And maybe there are other strange colours in other averages that we don't want in the mix, as well as some that we do. What matters is that you know what's in and what's out, and that you are sure you have achieved the mix you want. Averages are an abstraction, a useful one, but an abstraction all the same. If we look at them without knowing what it is that we have abstracted from, we will be misled. It is an average, but an average of what? Remember the vibrancy and variety of real life. Remember the rainbow.

6

Targets: The Whole Elephant

One number, because it implies one definition, is almost never enough. What single measure, for example, would you choose to express your life's worth? How about your salary? For some, that would do nicely. Most would feel cheapened. Or maybe longevity captures it? Up to a point, perhaps, but we might soon ask to what purpose you had put all those years. Whatever the measurement, if one part is taken for the whole, it can soon become comical.

Much social and political life is every bit as rich and subtle as our own, and just as resistant to caricature by a single objective defined with a single measurement. If you want to summarise like this, you have to accept the violence it can do to complexity.

This is why targets struggle. They necessarily try to glimpse a protean whole through the keyhole of a single number. So the strategy with targets is much like that for the average: think about what they do not measure, as well as what they do.

A consultant was asked to come to casualty urgently. She hurried down to find a patient about to breach the four-hour target for trolley waits. She also found another patient, who seemed in more pressing need, and challenged the order of priority. But no, they wanted her to leave that one: 'He's already breached.'

When we asked people to email *More or Less* with examples of their personal experience of 'gaming' in the NHS – or what is sometimes called 'hitting the target but missing the point' – we had little idea of the extent, or variety out there.

Here is another: 'I work in a specialist unit. We're always getting people sent to us by Accident and Emergency (A & E) because once they are referred the patient is dealt with as far as they are concerned and has met their target. Often, the patients could have been seen and treated perfectly well in A & E. But if they were sent to us, they might be further away from home and have to wait several days to be treated because they are not, and frankly never were, a priority for a specialist unit.'

And another: 'I used to work for a health authority where part of my job was to find reasons to reset the clock to zero on people's waiting times. We were allowed to do this if someone turned down the offer of an appointment, for example. But with the waiting-time targets, we began working much harder to find opportunities to do it, so that our waiting times would look shorter.'

Why do targets go wrong? The first reason has to do with their attempt to find one measurable thing to represent the whole. An analogous problem appears in the Indian fable of the Blind Men and the Elephant. The best-known Western version is by an American poet, John Godfrey Saxe (1816–87):

It was six men of Indostan
To learning much inclined,
Who went to see the Elephant
(Though all of them were blind),

That each by observation
Might satisfy his mind

But their conclusions depended entirely on which bit of the elephant they touched, so they decided, separately, that the elephant was like a wall (its side), a snake (trunk), a spear (tusk), a tree (leg), a fan (ear), or a rope (tail).

And so these men of Indostan
Disputed loud and long,
Each in his own opinion
Exceeding stiff and strong,
Though each was partly in the right,
And all were in the wrong!

To be partly in the right is often the best that targets can do, by showing only part of the picture. The ideal, obviously, is to show us the whole elephant, but it is no slight on numbers to say that they rarely can. Hospital waiting lists or ambulance response times are worth measuring but, even if the numbers are truthful – and we will see shortly why they are often not – they are inevitably grossly selective, and say nothing at all about the thing we care for most, which is the quality of care. Yes, patients were seen quickly, but did they die at the hands of inadequately trained personnel, whose job it was to arrive at the scene quickly in order to meet the target but who lacked the skills of a capable paramedic? Ambulance trusts really have done such things.

Targets, and their near allies performance indicators, face just such a dilemma. One measurement is expected to stand as the acid test, one number to account for a wide diversity of objectives and standards, while the rest ... out of sight, undefined, away from scrutiny, unseen through the keyhole, who cares about the rest?

Monomania, including mono-number mania, is potentially dodgy in any guise. The best strategy with targets, and indeed with any single-number summary, is to be clear not only what they measure, but what they do not, and to perceive how narrow the definition is. So that when a good health service is said to mean short waiting times, and so short waiting times are what is measured, someone might stop to ask: 'and is the treatment at the end *any good*?'

Quality of healthcare tends not to be measured because no one has worked out how to do it with any subtlety. So we are left with a proxy, the measurement that can be done rather than the one that ideally should be done, even though it might not tell us what we really want to know, a poor shadow that might look good even though quality is bad, or vice versa – right in part, and also wrong.

In October 2006 on *More or Less* we investigated behaviour around the four-hour target limit for trolley waits in casualty, and found evidence to suggest that hospitals were formally admitting people (often involving little more than a move down the corridor to a bed with a curtain), in some cases solely in order to stop them breaching the limit. That is, they were not really being treated inside the waiting target, it just looked that way. Sometimes this practice was probably clinically justified, and it was genuinely preferable for patients to be somewhere comfortable while investigations were completed. In other cases the patients were out again fifteen minutes after being admitted, feeding the suspicion that admittance had been a gaming strategy, not a clinical need. A massive increase in what are called 0-night-stay admissions added to the suspicion that this practice was so widespread as to be partly the result of playing the system to meet the target. While the number of people arriving at A & E between 1999/2000 and 2004/5 went up 20%, the number admitted went up 40%.

None of this is illegal; to some extent it is perfectly rational for those taking the decisions, under pressure in a busy A & E department, to respond to the incentives they are given. If those doctors and nurses simply cannot see everyone who has come to A & E in the four-hour limit, and will be penalised if they miss that target, admitting them to a ward seems the next best thing. That way, they will be seen (eventually), and the A & E department hits its four-hour target to see or admit. But the problem has recently worsened since the government has introduced a system of payment by results. Now, every time a hospital admits someone from A & E, the hospital is paid £500. So the gain from admitting someone from A & E is no longer just that it helps meet the four-hour target, but also raises funds for the hospital. More funds for the hospital means less somewhere else in the NHS, and if the admissions are in truth unjustified, implies a serious misallocation of resources.

Since our report, the Information Centre for Health and Social Care, NHS data-crunchers in chief, has said at a public conference that we caused it to launch its own investigation, despite Department of Health denials that there was any kind of problem to be investigated. A department official told us that the changes revealed by these statistics represented good clinical practice. If so, it is a practice the government does not want too much of, since the rise from year to year in 0-day admissions that it is prepared to pay for has now been capped – a strange way of showing approval. We have learned that one hospital has agreed to cut the fee it charges to primary care trusts for such admissions, and another is renegotiating.

According to Gwyn Bevan, Professor of Management Science at the London School of Economics, and Christopher Hood, Professor of Government and Fellow of All Souls College Oxford, the current faith in targets rests on two 'heroic' assumptions.

The first is the elephant problem: the parts chosen must robustly represent the whole, a characteristic they call 'synecdoche', a figure of speech borrowed from rhetoric. For example, we speak of hired hands, when we mean, of course, the whole worker. The second heroic assumption is that the design of the target will be 'game proof'.

The difficulties with the second follow hard on the first. Because it is tough to find the single number to stand for what we want out of a complex system delivering multiple objectives (just as it is in life), because one part seldom adequately speaks for all, it gives licence to all kinds of shenanigans in the parts that are untargeted, out of sight and often out of mind. So if your life did happen to be judged solely on your income, with no questions asked about how you came by it, and if there was no moral imperative to hold you back, 'gaming' would be too gentle a word for what you might get up to.

Bevan and Hood have catalogued a number of examples of targets in the health service that aimed at something which seemed good in itself, but which pulled the rug on something else. The target was hit, the point was missed, damage was done elsewhere.

In 2003, the Public Administration Select Committee found that the waiting-time target for new ophthalmology outpatient appointments had been achieved at one hospital by cancelling or delaying follow-up appointments. As a result, over two years at least twenty-five patients were judged to have gone blind.

In 2001 the government said all ambulances should reach a life-threatening emergency (category A) within eight minutes, and there was a sudden massive improvement in the number that did, or seemed to. But what is a 'life-threatening emergency'? The proportion of emergency calls logged as category A varied fivefold across trusts, ranging from fewer than 10 per cent to more than 50 per cent. It then turned out that some ambulance services were doctoring their response times – lying, to put it plainly. They cheated on the numbers, but it was also numbers that found them out, when it was discovered that there was a suspiciously dense concentration of responses recorded just inside eight minutes, causing a sudden spike in the graph, and almost no responses just above eight minutes. This was quite unlike the more rounded curve of response times that one would expect to see, and not a pattern that seemed credible. Figure 4 shows the pattern that led to suspicion that something was amiss. There was even some evidence that more urgent cases were sometimes made to wait behind less urgent (but still category A) cases, to meet the targets.

On waiting times in A & E, where there has been a sharp improvement in response to a target, but ample evidence of misreporting, Bevan and Hood concluded: 'We do not know the extent to which these were genuine or offset by gaming that resulted in reductions in performance that was not captured by targets.' Gwyn Bevan, who is, incidentally, a supporter of targets in principle and has even helped devise some, told us that when a manager worked honestly but failed to reach a target, then saw another gaming the system, hitting the target and being rewarded, the strong incentive next time would be to game like the rest of them: bad behaviour would drive out good.

And there are other examples in addition to those examined

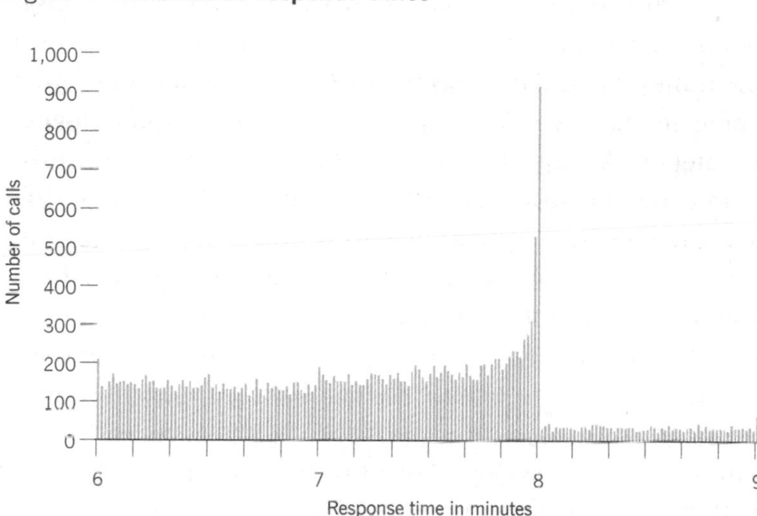

Figure 4 **Ambulance response times**

by Bevan and Hood. The new contract for GPs, which began in 2004, rewards them for, among other things, ensuring that they give patients at least ten minutes of their time (a good thing, if the patients need it). But that also gives an incentive to spin out some consultations that could safely be kept short, 'asking after Aunty Beryl and her cat', as one newspaper put it. The creditable aim was to ensure people received proper attention. The only measurement available was time, and so a subtle judgement was summarised with a single figure, that became a target, that created incentives, that led to suspicions of daft behaviour.

Then there is Britain's proud record for having the safest roads in Europe. We measure this by the number of accidents. But there are both elephant and gaming problems with our road-safety statistics. The first comes about because we define road safety by what happens on the roads. We take no account of the fact that many roads have become so fast, dual carriageways and the like, that pedestrians simply avoid them. Risk aversion is not the same as safety. So in some ways the roads might not be safer but more dangerous, even though casualties may be falling.

But are casualties falling? Over the very long term they are, unquestionably. More recent evidence for adult casualties is less clear. The government has targeted road accidents, telling police forces they would be judged by their success in reducing the number killed and seriously injured on the roads. By 2010, it says, there should be 40 per cent fewer accidents overall than in the baseline period of 1994–8. As the target was introduced, the numbers began to fall and the government hailed a dramatic success.

Then, in July 2006, the *British Medical Journal* reported an investigation of trends in accident statistics. This said that, according to the police, rates of people killed or seriously injured on the roads fell consistently from 85.9 accidents per 100,000 people in 1996 to 59.4 per 100,000 in 2004.

But the police are not the only source for such statistics and the *BMJ* authors' inspiration was to check them against hospital records. There, recorded rates for traffic injuries were almost unchanged at 90.0 in 1996 and 91.1 in 2004. The authors concluded that the overall fall seen in police statistics for non-fatal road traffic injuries 'probably represents a fall in completeness of reporting of these injuries'.

The police have some discretion about how they record an injury, and it is in this category of their statistics that the bulk of the improvement seems to have occurred. Deaths on the roads, where there is little scope for statistical discretion, have been largely flat in recent years in both police and hospital statistics. So, once targeted, it seems the police noted a fall in the one kind of accident over which they had some discretion in the reporting, but no other. And others didn't observe the same change. So did the accidents really go down, or did the police simply respond to the target by filling fewer notebooks?

One last example in a list long enough to suggest a generic problem. In response to concern about recycling rates in Britain lagging behind others in Europe, the government targeted them. Local councils responded with ingenuity and started collecting waste they had never tried to collect before, but that they could easily recycle. They called it, to give the whole enterprise a lick of environmental respectability, green waste. We do not really know what happened to this waste before: some was probably burnt, some thrown on the compost, some no doubt went in the black bin with the other rubbish. Now a big van came to collect it instead. Being heavy with water (waste is measured by weight), the vegetation did wonders for the recycling rate. There have even been stories of green waste being sprayed with water to make it even heavier. But is that really what people had in mind when they said there should be more recycling?

If all this leads to the conclusion that measurement is futile, then it is a conclusion too far: whether your income is $1 a day or $100 matters enormously. The key is to know the number's limitations: how much does it capture of what we really want to know? How much of the elephant does it show us? When is it wise to make this number everyone's objective? How will people behave if we do?

This suggestion – that targets and other summary indicators need to be used with humility – implies another; that until they are, we must treat them with care or even suspicion.

But little effort goes into checking that the huge volume of numbers generated by various audits, across the public sector, are reliable or that they haven't led to gaming. We could, for example, check the performance data against customer experience. In 2002/3, according to official figures, in 139 out of 158 acute trusts, 90 per cent of patients were seen at A & E in less than four hours. In a survey of patients, 69 per cent reported the same.

More of this is beginning to be done, in checks by the Healthcare Commission, for example, but it could go much further, to include more checks like the shrewd investigation of ambulance response times, but there is an incentive not to bother. Both the target setter – the government – and target managers want the numbers to look good, and critics allege collusion between them. Some of the early measurements of waiting times, for example, were a snapshot during a short period that – dubiously – was announced well in advance, giving hospitals a nod and a wink to do whatever was necessary in that period, but not others, to score well. It was known that hospitals were diverting resources during the measured period to hit the target, before moving them back again.

It is as well there is room for improvement, since Bevan and Hood say that targets and performance indicators are not easily dispensed with: alternatives like command and control from the centre are not much in favour either, nor is a free market in healthcare. If target setters were serious about the problem of gaming, an interesting approach would be to be a bit vague about the target, so that no one is quite sure how to duck it (will it be first appointments or follow ups?), introduce some randomness into the monitoring, and there could be more systematic monitoring of the integrity of the numbers.

They conclude: 'Corrective action is needed to reduce the risk of the target regime being so undermined by gaming that it degenerates, as happened in the Soviet Union.'

We haven't yet reached that point of ridicule, though it might not be far off. The Police Federation has complained (May 2007) that its members are spending increasing time prosecuting trivial crimes and neglecting more important duties simply in order to meet targets for arrest or summary fine. Recent reports include that of a boy in Manchester arrested under firearm laws for being in possession of a plastic pistol, a youth in Kent for throwing a slice of cucumber at another youngster, a man in Cheshire for being 'in possession of an egg with intent to throw', a student fined after 'insulting a police horse'. The absurdity of such fatuous activity – all in the name of improving police performance – is that it will give the appearance of an increasing crime rate.

Underlying many of the problems here is the simple fact that measurement is not passive; it influences what is measured. And many of the measurements we hear every day, if strained too far, may have caricatured the world and with that changed it in ways we never intended. Numbers are pure and true; counting never is. That limitation does not ruin counting by any means, but if you forget it, the world you think you know through numbers will be a neat and tidy illusion.

7
Risk: Make It Personal II

Numbers have amazing power to put life's anxieties into proportion: Will it be me? What happens if I do? What if I don't? Probabilities can *help.*

Yet this power is squandered through an often-needless misalignment of the way people habitually think, and the way risks and uncertainties are typically reported.

The news says, 'Risk up 42 per cent', a solitary, abstract number. All you want to know is, 'Does that mean me?' There you are, wrestling with your fears and dilemmas, and the best you've got to go on is a percentage, typically going up, and generally no help whatsoever.

Our thoughts about uncertainty are intensely personal, but the public and professional language can be absurdly abstract. No surprise, then, that when numbers are mixed with fear the result is often not the insight it could be, but confusion, and fear out of proportion.

It need not be like this. It is often easy to bring the numbers back into line with personal experience. Why that is not done is sometimes a shameful tale. When it is done, we often find that statements about risk that had appeared authoritative and somehow scientific were telling us nothing useful.

The answer to anxiety about numbers around risk and uncertainty is, like other answers here, simple: be practical and human.

'For every alcoholic drink a woman consumes, her risk of breast cancer rises by 6 per cent.'

That's pure garbage, by the way, despite its prominence on the BBC's national TV news bulletins on 12 November 2002, and it's soon obvious why: if true, every woman who drank regularly – and plenty who liked an occasional tipple – would be a certainty for breast cancer in time for Christmas. A 6 per cent increase with every glass soon adds up; about seven bottles of wine over a lifetime would make you a sure thing.

Not much in life is this certain and breast cancer certainly isn't. Still, ludicrous implausibility didn't stop the claim making the headlines. With luck, viewers were less easily taken in than the journalists who produced the report, since this is a passing piece of innumeracy that should have been easy to spot.

So what was the real meaning of the study so mangled in the news? It was true that research had shown a link between alcohol and breast cancer, but the first thing to do when faced with an increase in risk is to try to quell the fear and concentrate on the numbers. And the first question to ask about this number couldn't be simpler, that old favourite: 'how big is it?'

Cancer Research UK, which publicised the research – a perfectly professional piece of work, led by a team at Oxford University – announced the results as follows:

> A woman's risk of breast cancer increases by 6 per cent for every extra alcoholic drink consumed on a daily basis, the world's largest study of women's smoking and drinking behaviour reveals.

It added that for two drinks daily, the risk rises by 12 per cent. This was how most other news outlets reported the story. So there was something in this 6 per cent after all, but it was 6 per cent if you had one drink every day of your adult life, not 6 per cent for every single drink, a small change of wording for a not-so-subtle difference of meaning.

This at least is accurate, but also still meaningless. What's missing is how big the risk was to start with. Until we know this, being told only by how much it has changed is no help at all.

Why do we need answers to both? Imagine two people running a race. One is twice as fast as the other. 'Fred twice as fast as Eric,' says the headline. But is Fred actually any good (is the risk high)? If the difference between them is all you know, then you know next to nothing of Fred's ability. They might both be hopeless, with Fred plodding but Eric staggering, with a limp, and blisters, on no training and an eight-pint hangover. Or Eric might be a respectable runner and Fred a world record holder. If all we know is the relative difference (twice as good as the other one), and we don't know how good the other one was, then we're ignorant of most of what matters. Similarly, if the only information we have about two risks is the difference between them (one is 100 per cent more risky than the other, without even knowing the other), we know nothing useful about either.

It is plain to see that the same percentage rise in risk can lead to a very different number at the end depending on the number at the beginning. It is astonishing how often news reports do not tell us the number at the beginning, or at the end, but only the difference.

'Stop-and-search of Asians up fourfold'; 'Teenage pregnancy up 50 per cent in London borough'. But have the stop-and-search numbers for the area in question gone up from one person last quarter to four people this quarter, and plausibly within the sort of variation that might ordinarily be expected, or up from 100 to 400 and indicative of a dramatic and politically charged change in police tactics? Are the teenage pregnancy rates up from two last year to three this year, or from 2,000 to 3,000? Viewers of the TV quiz show *Who Wants to be a Millionaire?* know that doubling the money with the next correct answer varies hugely depending how much you have won already. 'Contestant doubles her money' tells us nothing.

So why does the news sometimes report risk with only one number, the difference between what a risk was and what it becomes? '*Risk for drinkers up 6 per cent!*' Six per cent of what? What was it before? What is it now? These reports, whatever their authors think they are saying, are numerical pulp.

How do we make the numbers for breast cancer meaningful? There is a formal way and the human way. First, the formal way, which you can ignore if necessary. We need to know the baseline risk – the risk of contracting the disease among women who do not drink. About 9 per cent of women will have breast cancer diagnosed by the time they are 80. Knowing this, we can get an idea of how much more serious it is for drinkers, and because the baseline risk is fairly small, an increase of 6 per cent in this risk will still leave it relatively small. Like the slow runner, a 6 per cent increase in speed will not make him a contender.

(To do the calculation properly, take the roughly 9 per cent risk we started with and work out what 6 per cent of that would be. Six per cent of 9 per cent is about 0.5 per cent. That is the additional risk of drinking one unit daily, a half of one per cent, or for two drinks daily about 1 per cent. But that is still not as intuitively easy to grasp as it could be. Many people often struggle to understand percentages in any form. In a survey, 1,000 people were asked what '40 per cent' meant: (a) one quarter, (b) 4 out of 10, or (c) every 40th person. About one third got it wrong. To ask, as we just did: 'what is 6 per cent of 9 per cent?' will surely baffle even more.)

Two days after the study made so many headlines Sarah Boseley, the *Guardian* newspaper's level-headed health editor, wrote an article entitled 'Half a Pint of Fear': 'How many of us poured a glass of wine or a stiff gin last night without wondering, if only briefly, whether we might be courting breast cancer? ... There will undoubtedly be women who turned teetotal instantly.'

Now why would they do that? Only, as she went on to suggest, because they had been panicked by the 6 per cent figure when in truth the difference a drink a day made could have, and should have, been presented in a less alarming light. That is, unless creating alarm was the whole point.

Our interest is not advocacy, but clarity. So let us start again without a practice that makes the risk appear as big as possible. Let's do away with percentages altogether and speak, as journalists should, of people.

Here is a simpler way to describe what the reports horribly failed to do, looking at the effect of two drinks a day, rather than one, to keep the numbers round.

'In every 100 women, about nine will typically get breast cancer in a lifetime. If they all had two extra drinks every day, about ten would.'

And that's it.

All the information so mangled or neglected by talk only of the increase in relative risk, or percentages of percentages, is contained clearly in those two sentences. You can quickly see that in a hundred women having two alcoholic drinks every day there would be about one extra case of cancer.

Though 1 in 100 is a small proportion, because the British population is large, this would still add up to quite a few cases of breast cancer. Our point is not to make light of a frightening illness, nor to suggest that the risk of cancer is safely ignored. It is because cancer is frightening that it is important to make sense of the risks in a way most people can understand. Otherwise we are all at the mercy of news reports which appear like bad neighbours, leaning over the fence as we go about our lives and saying with a sharp intake of breath: 'You don't want to be doing that.' And maybe you don't. But let us take the decision on the basis of the numbers presented in a way that makes intuitive human sense.

The number of people affected in every hundred is known by statisticians as a natural frequency. It is not far from being a percentage, but is a little less abstract, and that helps. For a start, it is how people normally count, so it feels more intuitively intelligible. It also makes it much harder to talk about relative differences, harder to get into the swamp of talking about one percentage of another. By the way, if you think we exaggerate the difficulty for many people of interpreting percentages, we will see in a moment how even physicians who have been trained in medical statistics make equally shocking and unnecessary mistakes when interpreting percentages for their patients' test results.

Natural frequencies could easily be adopted more widely, but are not, so tempting the conclusion that there is a vested interest both for advocacy groups and journalists in obscurity. When the information could be conveyed in so plain a fashion, why do both often prefer to talk about relative percentage risks without mentioning the absolute risk, all in the most abstract terms? The suspicion must be that this allows the use of 'bigger' numbers ('six' per cent is big enough perhaps to be a scare, the absolute change 'half of one per cent', or even 'one woman in every 200' is probably less disturbing). Bigger numbers win research grants and sell causes, as well as newspapers.

One standard defence to this is that no one is actually lying. That is true, but we would hope for a higher aspiration from cancer charities or serious newspapers and broadcasters than simply getting away with it. The reluctance to let clarity get in the way of a good story seems particularly true of health risks, and isn't only a habit of some journalists. There are, in fact, international guidelines on the use of statistics that warn against the use of unsupported relative risk figures. Cancer Research UK in this case seems either to have been unaware of, or to have ignored, these guidelines in this case in favour of a punchier press release. When we have talked to journalists who have attended formal training courses in the UK, not one has received any guidance about the use of relative risk figures.

In January 2005 the president of the British Radiological Protection Board announced that risks revealed in new medical research into mobile phones meant children should avoid them. The resulting headlines were shrill and predictable.

He issued his advice in the light of a paper from the Karolinska Institute in Sweden that suggested long-term use of mobiles was associated with a higher risk of a brain tumour known as an acoustic neuroma. But how big was the risk? The news reports said that mobile phones caused it to double.

Once again, almost no one reported the baseline risk, or did the intuitively human thing and counted the number of cases implied by that risk; the one honourable exception we found – all national newspapers and TV news programmes included – being a single story on BBC News Online. A doubling of risk sounds serious, and it might be. But as with our two men in a race, it could be that twice as big, just like twice as good, or twice as bad, doesn't add up to much in the end.

With mobile phones you could begin with the reassurance that these tumours are not cancerous. They grow, but only sometimes, and often slowly or not at all after reaching a certain size. The one problem they pose if they do keep growing is that they put pressure on surrounding brain tissue, or the auditory nerve, and might need cutting out. You might also note that the research suggested these tumours did not appear until after at least ten years of at least moderate mobile phone use. And you might add to all those qualifications the second essential question about risk: how big *was* it? That is, what was the baseline?

When we spoke to Maria Feychting of the Karolinska Institute, one of the original researchers, a couple of days after the story broke, she told us that the baseline risk was 0.001 per cent or, expressed as a natural frequency, i.e. people, about 1 in 100,000. This is how many would ordinarily have an acoustic neuroma if they didn't use a mobile phone. With ten years regular phone use, the much-reported doubling took this to 0.002 per cent, or 2 people in every 100,000 (though it was higher again if you measured by the ear normally nearer the phone). So regular mobile phone use might cause the tumours in 0.001 per cent of those who used them, or one extra person in every 100,000 in that group.

Would Maria Feychting stop her own children using mobile phones? Not at all: she would rather know where they were and be able to call them. She warned that the results were provisional, the study small, and quite different results might emerge once they looked at a larger sample. In fact, it was usually the case, she said, that apparent risks like this seemed to diminish with more evidence and bigger surveys. Two years later the worldwide research group looking into the health effects of mobile phones – Interphone – of which the Karolinska Institute team was a part, did indeed produce another report drawing on new results from a much larger sample. It now said there was no evidence of increased risk of acoustic neuroma from mobile phones, the evidence in the earlier study having been a statistical fluke, the product of chance, which in the larger study disappeared.

A great many percentage rises or falls, in health statistics, crime, accident rates and elsewhere, are susceptible to the same problem and the same solution. So we can make this the standard response to any reported risk: imagine that it is the difference between those two men in a race. When you are told that one risk/runner is – shock! – far ahead of the other, more than, bigger than the other, always ask: but what's the other one/risk like? Do not only tell me the difference between them. Better still, reports of risks could stick to counting people, as people instinctively do, keep percentages to a minimum and use natural frequencies. Press officers could be encouraged to do the same, and then we could all ask: how many *extra* people per 100, or per 1000 might this risk affect?

Risk is one side of uncertainty. There is another, potentially as confusing and every bit as personal.

Imagine that you are a hardworking citizen on the edge, wide-eyed, sleep-deprived, while nearby in the moonlight someone you can barely now speak of in civil terms has a faulty car alarm. It's all right; we understand how you feel, even if we must stop short, naturally, of condoning your actions as you decide that if he doesn't sort out the problem, right now, you might just remember where you put the baseball bat.

The alarm tells you with shrill self-belief that the car is being broken into; you know from weary experience that the alarm in question can't tell theft from moped turbulence. You hear it as a final cue to righteous revenge; a statistician, on the other hand, hears a false positive.

False positives are results that tell you something important is afoot, but are wrong. The test was done, the result came in, it said 'yes', but mistakenly, for the truth was 'no'. All tests have a risk of producing false positives.

There is also a risk of error in the other direction, the false negative. False negatives are when the result says 'no', but the truth is 'yes'. You wake up to find, after an uninterrupted night's sleep, that the car has finally been stolen. The alarm – serves him right – had nothing to say about it.

There are a hundred and one varieties of false positive and negative; the cancer clusters in Chapter 4 on chance are probably a false positive, and it is in health generally that they tend to be a problem, when people have tests and the results come back saying they either have or haven't got one condition or another. Some of those test results will be wrong. The accuracy of the test is usually expressed as a percentage: 'The test is 90 per cent reliable'. It has been found that doctors, no less than patients, are often hopelessly confused when it comes to interpreting what this means in human terms.

Gerd Gigerenzer is a psychologist, Director of the Centre for Adaptive Behaviour and Cognition at the Max Planck Institute in Berlin. He asked a group of physicians to tell him the chance of a patient truly having a condition (breast cancer) when a test (a mammogram) that was 90 per cent accurate at spotting those who had it, and 93 per cent accurate at spotting those who did not, came back positive.

He added one other important piece of information: that the condition affected about 0.8 per cent of the population for the group of 40–50-year-old women being tested. Of the twenty-four physicians to whom he gave this information, just two worked out correctly the chance of the patient really having the condition. Two others were nearly right, but for the wrong reasons. Most were not only wrong, but hopelessly wrong. Percentages confused the experts like everyone else.

Quite a few assumed that, since the test was 90 per cent accurate, a positive result meant a 90 per cent chance of having the condition, but there was a wide variety of opinion. Gigerenzer comments: 'If you were a patient, you would be justifiably alarmed by this diversity.' In fact, more than nine out of ten positive tests under these assumptions are false positives, and the patient is in the clear.

To see why, look at the question again, this time expressed in terms that make more human sense, natural frequencies.

Imagine 1,000 women. Typically, eight have cancer, for whom the test, a fairly but not perfectly accurate test, comes back positive in 7 cases. The remaining 992 do not have cancer, but remember that the test can be inaccurate for them too. Nearly 70 of them will also have a positive result. These are the false positives, people with positive results that are wrong.

Now we can see easily that there will be about 77 positive results in total (the true positives and the false positives combined) but that only about 7 of them will be accurate. This means that for any one woman with a positive test, the chance that it is accurate is low and not, as most physicians thought, high.

The consequences, as Gigerenzer points out, are far from

trivial: emotional distress, financial cost, further investigation, biopsy, even, for an unlucky few, unnecessary mastectomy. Gigerenzer argues that at least some of this is due to a false confidence in the degree of certainty conferred by a test that is at least 90 per cent accurate. If positive tests were reported to patients with a better sense of their fallibility, it would ease at least some of the emotional distress. But how does that false confidence arise? In part, because the numbers used are not in tune with human instinct to count.

Uncertainty is a fact of life. Numbers, often being precise, are sometimes used as if they overcame it. A vital principle to establish is that many numbers will be uncertain, and we should not hold that against them. Even 90 per cent accuracy might imply more uncertainty than you would expect. The human lesson here is that since life is not certain, and since we know this from experience, we should not expect numbers to be any different. They can clarify uncertainty, if used carefully, but they cannot beat it.

Having tried to curb the habit of over-interpretation, we need to restrain its opposite, the temptation to throw out all such numbers. Being fallible does not make numbers useless, and the fact that most of the positives are false positives does not mean the test is no good. It has at least narrowed the odds, even if with nothing like 90 per cent certainty. Those who are positive are still unlikely to have breast cancer, but they are a little more likely than before they were tested. Those who are negative are now even less likely to have it than before they were tested.

So it is not that uncertainty means absolute ignorance, nor that the numbers offer certainty, rather that they can narrow the scope of our ignorance. This partial foretelling of fate is an extraordinary achievement. But we need to keep it in proportion, and we certainly need to get the right way round the degree of risk, whether likely or not. The overwhelming evidence is that we are more likely to judge this correctly if we use natural frequencies when we can and count people, as people do, rather than use percentages.

What generally matters is not whether a number is right or wrong, they are often wrong, but whether numbers are so wrong as to be misleading. It is standard practice among statisticians to say how wrong they think their numbers might be, though we might not even know in which direction – whether too high or too low. Putting an estimate on the potential size of the error, which is customarily done by saying how big the range of estimates needs to be before we can be 95 per cent sure it covers the right answer (known as a confidence interval), is the best we can do by way of practical precaution against the number being bad. Though even with a confidence interval of 95 per cent there is still a 5 per cent chance of being wrong. This is a kind of modesty the media often ignore. The news often doesn't have time, or think it important, to tell you that there was a wide range of plausible estimates and that this was just one, from somewhere near the middle. So we don't know, half the time, whether the number is the kind to miss a barn door, enjoying no one's confidence, or if it is a number strongly believed to hit the mark.

We accuse statisticians of being overly reductive and turning the world into numbers, but statisticians know well enough how approximate and fallible their numbers are. It is the rest of us who perform the worst reductionism whenever we pretend the numbers give us excessive certainty. Any journalist who acts as if the range of uncertainty does not matter, and reports only one number in place of a spread of doubt, conspires in a foolish delusion for which no self-respecting statistician would ever fall. Statistics is an exercise in coping with, and trying to make sense of, uncertainty, not in producing certainty. It is usually frank in admission of its doubt and we should be more willing to do the same.

If ever you find yourself asking, as you contemplate a number, 'How can they be so precise?' the answer is that they probably can't, and probably weren't, but the reporting swept doubt under the carpet in the interests of brevity. If, somewhere along the line, the uncertainty has dropped out of a report, it will probably pay to find out what it had to say.

If we accept that numbers are not fortune tellers and will never tell you everything, but can tell you something, then they retain an astonishing power to put a probability on your fate. The presentation of the numbers might have left a lot to be desired, but the very fact that we can know – approximately – what effect drinking regularly will have on your chances of breast cancer, is remarkable. Picking out the effect of alcohol from all the other lifetime influences on your health is a prodigious undertaking and the medical surveys that make it possible are massive data-crunching exercises. Having gone to all that effort, it is a scandal not to put it to proper use and hear clearly what the numbers have, even if with modesty, to say.

8

Sampling: Drinking from a Fire Hose

Seek numbers' origins and you find something surprising. Many of the hundreds printed and broadcast every day have routinely, necessarily, skimped on the job.

To know what they are worth, you need to know how they are gathered. But few know the shortcuts taken to produce even respected numbers: the size of the economy or trade, the profit companies make, how much travel and tourism there is, UK productivity, the rate of inflation, the level of employment ... as well as controversial numbers like Iraq war dead, HIV/Aids cases, migrants, and more.

Their ups and downs are the bread and butter of news, but none is a proper tally. Instead, only a few of each are counted, assuming they are representative of the rest, then multiplied to the right size for the whole country.

This is the sample, the essence of a million statistics, like the poet's drop of water containing an image of the world in miniature – we hope. But the few that are counted must *mirror the others or the whole endeavour fails; so which few? Choose badly and the sample is skewed, the mirror flawed, and for a great many of the basic facts about us, our country and our economy, only error is multiplied.*

A huddle of officials gathers at Dover docks in the cold early morning. Their job is to count, and much is at stake: what they jot on clipboards will lead others to bold conclusions, some saying we need more workers if Britain is to prosper, some that this imperils the British way of life. They count, or rather sample, migrants.

All the public normally hears and knows of the statistical bureaucracy at our borders is the single number that hits the headlines: in 2005 net immigration of about 180,000, or about 500 a day. This is sometimes enriched with a little detail about where they came from, how old they are, whether single or with children, and so on.

Officials from the International Passenger Survey, on the other hand, experience a daily encounter with compromising reality in which the supposedly simple process of counting people to produce those numbers is seen for what it is: mushiness personified. Here they know well the softness of people, people on the move even more so. What's more, they sample but a tiny fraction of the total.

The migration number for sea crossings begins when, in matching blue blazers on a dismal grey day, survey teams flit across the Channel weaving between passengers on the ferries, from the duty-free counter to the truckers' showers, trying in all modesty to map the immense complexities of how and why people cross borders. The problem is that to know whether passengers are coming or going for good, on a holiday or a booze cruise, or a gap year starting in Calais and aiming for Rio, there's little alternative to asking them directly. And so the tides of people swilling about the world, seeking new lives and fleeing old, heading for work, marriage or retirement in the sun, whether 'swamping' Britain or 'brain-draining' from it, however you interpret the figures, are captured for the record if they travel by sea when skulking by slot machines, half-way though a croissant, or off to the ladies' loo.

'Oi! You in the chinos! Yes, by the lifeboat! Where are you going?' And so they discover the shifting sands of human migration and ambition.

Or maybe not, not least because it is a lot harder than that. To begin with, the members of the International Passenger Survey teams are powerless – no one has to answer their questions – and, of course, they are impeccably polite; no job this for the rude or impatient. Next, they cannot count and question everyone, there is not enough time, so they take a sample which, to avoid the risk of picking twenty people drinking their way through the same stag weekend and concluding the whole shipload is doing the same, has to be as random as possible.

So, shortly before departure, they stand at the top of the eight flights of stairs and various lifts onto the passenger deck as the passengers come aboard, scribbling a description of every tenth in the file; of the rucksacked, the refugees, the suited or the carefree, hoping to pick them out later for a gentle interrogation: tall-bloke, beard, 'surfers do it standing up' T-shirt. That one is easy enough, not much likelihood of a change of clothes either, which sometimes puts a spanner in the works.

When several hundred Boy Scouts came aboard en route to the World Scouting Jamboree, the survey team assumed, with relief, that the way to tell them apart would be by the colour of their woggles, until it turned out that the Jamboree had issued everyone with new, identical ones. The fact that they were all going to the same place and all coming back again did not absolve the survey teams of the obligation to ask. The risks of woggle fatigue are an occupational hazard of all kinds of counting.

The ferry heaves into its journey and, equipped with their passenger vignettes, the survey team members also set off, like Attenboroughs in the undergrowth, to track down their prey, and hope they all speak English.

'I'm looking for a large lady with a red paisley skirt and blue scarf. I think that's her in the bar ordering a gin and lime.'

She is spotted – and with dignified haste the quarry is cornered: 'Excuse me, I'm conducting the International Passenger Survey. Would you be willing to answer a few questions?'

'Of course.' Or perhaps, with less courtesy: 'Nah! Too busy, luv.' And with that the emigration of an eminent city financier is missed, possibly. About 7 per cent refuse to answer. Some report their honest intention to come or go for good, then change their minds, and flee the weather, or the food, after three months.

The International Passenger Survey teams interview around 300,000 people a year, on boats and at airports. In 2005 about six or seven hundred of these were migrants, a tiny proportion from which to estimate a total flow in both directions of hundreds of thousands of people (though recently they began supplementing the routine survey data with additional surveys at Heathrow and Gatwick designed particularly to identify migrants). The system has been described by Mervyn King, Governor of the Bank of England – who is in charge of setting interest rates and has good reason for wanting to know the size of the workforce – as hopelessly inadequate.

This is counting in the real world. It is not a science, it is not precise, in some ways it is almost, with respect to those who do it, absurd. Get down to the grit of the way data is gathered, and you often find something slightly disturbing: human mess and muddle, luck and judgement, and always a margin of error in gathering only a small slice of the true total, from which hopeful sample we simply extrapolate. No matter how conscientious the counters, the counted have a habit of being downright inconvenient, in almost every important area of life. The assumption that the things we have counted are correctly counted is rarely true, cannot, in fact, often be true, and as a result our grip on the world through numbers is far feebler than we like to think.

That does not make it a waste of time. Flawed as the data is, it is usually better than no data at all; inadequate it may be, but improvement is not always easy without huge expense. The key point is to realise these numbers' implicit uncertainty and treat them with the care – not cynicism – they deserve. Few people realise how much of our public data comes from samples. A glance at National Statistics, the home page for data about anything to do with the economy, population or society, reveals a mass of information, nearly all based on samples, and only one of its featured statistics based on a full count:

> This year's official babies' names figures show that it's all change for the girls with Olivia and Grace moving up to join Jessica in the top three. Jack, Thomas and Joshua continue to be the three most popular boy's names, following the trend of previous years.

Names are fully recorded and counted; so too is the approximate figure for government spending and receipts. But on a day taken at random, the day of writing, everything else listed there – and there are a dozen of the most basic economic and social statistics – is based on a sample. The size of that sample for many routine figures, inflation or the effect of changes in tax or benefits, for example, is about 7,500 households, or about 1/3,000th of the actual number of households in the UK.

This is inevitable. It could take a lifetime to number nothing more than one life's events, so much is there to count, so little that is practicably countable. Expensive, inconvenient, overwhelming, the effort to quantify even the important facts would be futile, were it not for sampling. But sampling has implicit dangers, and we need to know both how widespread it is as a source of everyday numbers, and how it goes wrong.

The National Hedgehog Survey began in 2001. Less cooperative even than the Census-shy public, hedgehogs keep their heads down. In the wild there is no reliable way to count them short of rounding them up, dead or alive, or ripping through habitats, ringing legs – which would somewhat miss the point.

It is said hedgehogs are in decline. How do we know? We could ask gamekeepers how many they have seen lately – up or down on last year – and we could put the same questions in general opinion surveys. But the answers would be impressionistic, anecdotal, and we want more objective data than that. What do we do?

In 2002 the Hedgehog Survey was broadened to become the Mammals on Roads Survey. Its name suggests, as young hedgehog-lovers might say, an icky solution. The survey is in June, July and August each year, when mammals are on the move and when, to take the measure of their living population, we count how many are silhouetted on the tarmac. The more hedgehogs there are in the wild, so the logic goes, the more will snuffle to their doom on the bypass. In a small hog population this will be rare, in a large one more common.

But spot the flaw in the method: does it count hedgehogs or traffic density? Even if the hedgehog population was stable, more cars would produce more squashes. Or, as the more ingenious listeners to *More or Less* suggested, does the decline in road-kill chronicle the evolution of a smarter, traffic-savvy hedgehog which, instead of rolling into a ball at sight or sound of hazard, now legs it, and lives to snuffle another day beyond the ken of our survey teams? Or maybe climate change has led to alterations in the hedgehog life cycle through the year, which reduces the chance of being on the roads in the three monitored months.

The most recent Mammals on Roads Survey tells us that average counts in both England and Wales have fallen each year of the survey, and the numbers in Scotland last year are markedly lower than those at the start of the survey. In England, the decline is biggest in the Eastern and East Midlands region and in the South West; no one knows why.

The lesson of the survey is that we often go to bizarre but still unsatisfactory lengths to gather enough information for a reasonable answer to our questions, but that it will remain vulnerable to misinterpretation, as well as to being a poor sample. Yet, for all the potential flaws, this really is the best we can do in the circumstances. Most of the time we don't stop to think, 'How did they produce that number?' We have been spoiled by the ready availability of data into thinking it is easily gathered. It is anything but. Do not assume there is an obvious method bringing us a sure answer. We rarely know the whole answer, so we look for a way of knowing part, and then trust to deduction and guesswork. We take a small sample of a much larger whole, gushing with potential data, and hope: we try to drink from a fire hose.

In fact, drinking from a fire hose is relatively easy compared with statistical sampling. The problem is how you can be sure that the little you are able to swallow is like the rest. If it's all water, fine, but what if every drop is in some way different?

Statistical sampling struggles heroically to swallow an accurate representation of the whole, and often fails. What if the sample of flattened hedgehogs comes entirely from a doomed minority, out-evolved and now massively outnumbered by runaway hedgehogs? If so, we would have a biased count – of rollers (in decline) instead of runners (thriving). Unlikely, no doubt, and completely hypothetical, but how would we know for sure?

But such systematic bias in sampling is not only theoretical. It applies, for example, to a number that is capable from time to time of dominating the headlines: the growth of the UK economy.

The Bank of England, the Treasury, politicians, the whole business community and the leagues of economic analysts and commentators generally accept the authority of the figures for UK economic growth, compiled in good faith with rigorous and scrupulous determination by National Statistics. It is a figure that has governments trembling, the bedrock of every economic forecast, the measure of our success as an economy, the record of rising prosperity or recession.

It is also based on a sample. What's more, there is good evidence the sample is systematically biased in the UK against those very parts of the economy likely to grow fastest. One consequence has been that for the last ten years we have believed ourselves under-performing compared with the United States, when in fact we might have been doing better, not worse.

In the UK, it is hard to spot the growth of new start-up

businesses until we have seen their tax returns. That is long after the event, perhaps two years after the first estimates of GDP growth are published, before the growth of new businesses has been incorporated in the official figures. So the initial GDP figures fail to count the one area of the economy that is plausibly growing fastest – the new firms with the new ideas, creating new markets. On past form this has led frequently to initial under-reporting of economic growth in the UK by about half a percentage point. When growth moves along at about 2.5 per cent a year, that is a big error.

But it is neither perverse, nor incompetent, and arguably unavoidable. One alternative, in the short-term absence of hard data, is to guess. How fast do we think new parts of the economy *might be* growing? And having watched it happen for a while, we might reckon that half a per cent of GDP would be a decent stab at the right number, but it would essentially be a guess that past performance would continue unchanged, a guess with risks, as any investor knows. The view of Britain's National Statistician is that it is right to count what we can reasonably know, not what we might speculate about, but this does mean an undercount is likely (though not inevitable).

Guessing is what they do in the United States – and then tend to have to revise their estimates at a later date when real data is available, usually down, often significantly. By the time a more accurate figure comes out, of course, it is too late to make much impression on public perception because everyone has already gone away thinking the US a hare and the UK a sloth. Not on this data, it's not. Our opinions have been shaped by our sampling, and our sampling is consistently biased in a way that is hard to remedy accurately until later, by which time no one is interested any more in the figures for events up to two years old. In 2004 we thought UK growth was 2.2 per cent (not bad, but not great) compared to US growth of 3.2 per cent (impressive). After later revisions, UK growth turned out to be 2.7 per cent, and US growth also 2.7 per cent. The latest figures available, for the pace at which the US economy grew in the last three months of 2006, were revised down from an initial estimate of 3.5 per cent to 2.2 per cent by March 2007, a massive correction.

But what is the alternative? One would be to measure the whole lot, count every bit of business activity, at the moment of each transaction. Well, we could, just about, if we were willing to pay for and endure so much statistical bureaucracy. We already measure some of it this way, but in practice some dainty picking and choosing is inevitable.

Life as a fire hose, and samplers with tea cups and crooked fingers, is an unequal statistical fight. In truth, it is amazing we capture as much as we do, as accurately as we do. But we do best of all if we also recognise the limitations, and learn to spot occasions when our sample is most likely to miss something.

HIV/Aids cases are a global emergency, fundamentally impossible to count fully. The number of cases worldwide was estimated by UNAids (the international agency with responsibility for the figures – and for coordination of the effort to tackle them) at 40 million people in 2001. The number is believed to have risen ever since and now stands (according to UNAids' 2006 report) at 38 million. That's right, there is no mistake, the 'increase' is from 40 to 38 million: the rising trend has gone down, though there are wide margins of uncertainty around these figures.

The explanation for this paradox is in the sampling. Many of the numbers from the earlier surveys came from samples of the population taken at urban maternity clinics. There are at least two possible biases in this sample: one, anyone coming to a maternity clinic has most likely had sex – unprotected sex; two, urban areas might well have higher rates of HIV/Aids than rural areas.

UNAids believes its earlier estimates were too high, and has revised them in the light of other surveys. There never were, it now thinks, as many cases as there are now. So can we trust its new sampling methodology? The figures are contested, in both directions, some believing them too low, others still too high, and arguments will always rage. All we can do is keep exercising our imagination for ways in which the sample might be a warped mirror of reality.

It is worth saying in passing that UNAids thinks the problem has peaked in most places, adding that one growing contribution to the numbers is a higher survival rate. Not all increases in the numbers suffering from an illness are unwelcome: sometimes they represent people who in the past would have died, but are now alive and that is why there are more of them.

Still, doubts about accuracy or what the numbers represent do not discredit the conclusion that we face a humanitarian disaster: 2 million people are believed by UNAids to have died of HIV/Aids last year, and it is thought there were 4 million new cases worldwide. In some places, earlier progress in reducing the figures seems to be in reverse. These figures, though certainly still wrong, one way or another, would have to be very wrong indeed to be anything less than horrifying.

Reports of the figures tend to pick out a single number. But sampling is often so tricky that we put large margins of uncertainty around its results. The latest UN figures do not state that 38 million is *the* number; they say that the right number is probably somewhere between about 30 million and about

48 million, with the most likely number somewhere near the middle. One of these figures, note, is 60 per cent higher than the other. This is a range of uncertainty that news reporting usually judges unimportant: one figure will do, it tends to suggest. Where sampling gives rise to such uncertainty, one figure will seldom do.

Currently more controversial even than the figures for HIV/Aids is the Iraq war. Iraq has a population less than half the size of Britain's. In October 2006 it was estimated in research by a team from Johns Hopkins University, published in the *Lancet*, that nearly twice as many people had died as a result of the Anglo-American invasion/liberation of Iraq as Britain had lost in the Second World War, around 650,000 in Iraq to about 350,000 British war dead, civilian and combatants combined; that is, nearly twice as many dead in a country with a population (about 27 million), not much more than half of Britain's in 1940 (48 million).

Of those 650,000 deaths, about 600,000 were thought to be directly due to violence, a figure which was the central estimate in a range from about 400,000 to about 800,000. The comparison with the Second World War shows that these are all extremely big numbers. The political impact of the estimate matched the numbers for size, and its accuracy was fiercely contested.

It was, of course, based on a sample. Two survey teams visited fifty randomly selected sites of about forty households each, a total of 1,849 households with an average of about seven members (nearly 13,000 people). One person in each household was asked about deaths in the fourteen months prior to the invasion and in the period after. They asked in about 90 per cent of cases of reported deaths to see death certificates, and were usually obliged.

The numbers were much higher than those of the Iraq Body Count (which had recorded about 60,000 at that time), an organisation that uses two separate media reports of a war death before adding to its tally (this is a genuine count, not a sample), and is scrupulous in trying to put names to numbers. But because it is a passive count, it is highly likely to be (and is, by the admission of those who do it) an undercount.

But was it likely to have been such a severe undercount that it captured only about 10 per cent of the true figure? Attention turned to the sampling methodology of the bigger number. Because it was a sample, each death recorded had to be multiplied by roughly 2,200 to derive a figure for the whole of Iraq. So if the sample was biased in any way – if, to take the extreme example, there had been a bloody feud in one tiny, isolated and violent neighbourhood in Iraq, causing 300 deaths, and if all these deaths had been in one of the areas surveyed, but there had been not a single death anywhere else in Iraq, the survey would still produce a final figure of 650,000, when the true figure was 300.

Of course, the sample was nothing like as lopsided in this way, to this degree, but was it lopsided in some other? Did it, as critics suggested, err in sampling too many houses close to the main streets where the bombs and killings were more common, not enough in quieter, rural areas? Was there any manner in which the survey managed the equivalent of counting Dresden but missing rural Bavaria in a sample of Germany's war dead?

In our view, *if* the Iraq survey produced a misleading number (it is bound to be 'wrong' in the sense of not being precisely right, the greater fault is to be 'misleading'), it is more likely because of the kind of problem discussed in the next chapter – to do with data quality – than a bad sample.

What statisticians call the 'design' of the sample was in no obvious way stupid or badly flawed. But it is perfectly proper to probe that design, or indeed any sample, for weakness or bias.

To do this well, what's needed above all is imagination; that and enough patience to look at the detail. What kind of bias might have crept in? What are the peculiarities of people that mean the few we happen to count will be unrepresentative of the rest?

Even data that seems to describe us personally is prone to bias. You know your own baby, surely. But do you know if the little one is growing properly? The simple but annoying answer is that it depends what 'properly' means. In the UK, this is defined by a booklet showing charts that plot the baby's height, weight and head circumference, plus a health visitor sometimes overly insistent about the proper place on the chart for your child to be. The fact that there is variation in all children – some tall, some short, some heavy, some light – is small consolation to the fretful parent of many a child below the central line on the graph, the 50th percentile. In itself, such a worry is usually groundless, even if it is encouraged. All the 50th percentile means is that half of babies will grow faster and half slower. It is not a target, or worse, an exam that children pass or fail.

But there is a further problem. Who says this is how fast babies grow? On what evidence? The evidence naturally, of a sample. Who is in that sample? A variety of babies past that were supposed to represent the full range of human experience. Is there anything wrong with that?

Yes, as it happens. According to the World Health Organisation, not every kind of baby should have a place in the sample.

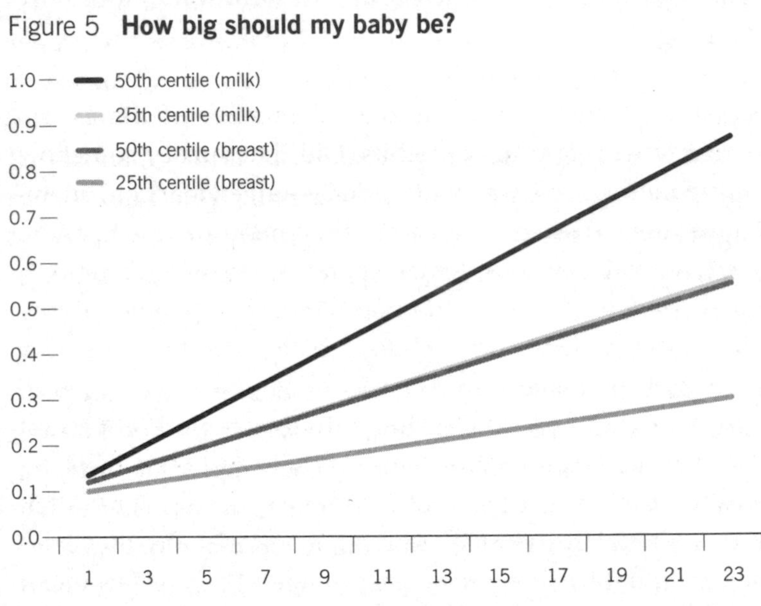

Figure 5 **How big should my baby be?**

The WHO wants babies breast-fed and says the bottle-fed should be excluded. This matters because bottle-fed babies tend to grow a little more quickly than breast-fed babies, to the extent that after two years the breast-fed baby born on the 50th percentile line will have fallen nearly to the bottom 25 per cent of the current charts (see Figure 5).

If the charts were revised, as the WHO wants, those formerly average-weight, bottle-fed babies would now move above the 50th percentile line, and begin to look a little on the heavy side, and the breast-fed babies would take their place in the middle.

So the WHO thinks the charts should set a norm – asserting the value of one feeding routine over another, and has picked a sample accordingly. This is a justifiable bias, it says, against bad practice, but a bias nevertheless, taking the chart away from description, and towards prescription simply by changing the sample.

Depending what is in a sample, all manner of things can change what it appears to tell us. Did that survey somehow pick up more people who were older, younger, married, unemployed, taller, richer, fatter; were they more or less likely to be smokers, car drivers, female, parents, left-wing, religious, sporty, paranoid … than the population as a whole, or any other of the million and one characteristics that distinguish us, and might just make a difference?

In one famous case it was found that a sample of Democrat voters in the United States were less satisfied with their sex lives than Republicans, until someone remembered that women generally reported less satisfaction with their sex lives than men, and that more women tended to vote Democrat than Republican.

Bias can seep into a population sample in as many ways as people are different, in their beliefs, habits, lifestyles, history, biology. A magazine surveys its readers and claims 70 per cent of Britons believe in fairies, but the magazine is called *Paranormal and Star-Gazers Monthly*. Simply by buying it, the readership proves itself already more predisposed to believe than the general population of 'Britons' to whom its conclusions are too readily applied.

Bias of some kind, intentional, careless, or accidental, is par for the course in many samples that go into magazine surveys, or surveys which serve as a marketing gimmick, and generally get the answer they want. For a quick sample of this practice – representative or not, we couldn't possibly say – try the following, and put your imagination to the test thinking of possible bias. All of these arrived on the desk of Rory Cellan

Jones, the BBC's business correspondent, during a week or two in the summer of 2006.

- New mothers spend £400 on average on toddlers' wardrobes, a survey says.
- Tea is the number one night-time drink for men and women.
- Research shows that 52 per cent of men in the city admit to wearing odd socks at least once a week (survey by an online sock retailer).
- 60 per cent of women would prefer to see celebrities looking slightly flawed, while 76 per cent of men in the UK prefer to see images of celebrities looking perfect (a survey courtesy of a make-up company and a high-definition TV channel).
- More than 20 million British homeowners have spent more than £150bn in tasteless home improvements that have reduced the value of their homes (a home insurance firm courteously tells us).

Bias is less likely in carefully designed surveys which aim for a random sample of the population and have asked more than the first half-dozen people that come along. But surveys asking daft questions and getting daft answers are not alone in finding that potential bias is lurking everywhere, with menace, trying to get a foot in the door, threatening to wreck the integrity of our conclusions.

Once more, it is tempting to give up on all things sampled as fatally biased and terminally useless. But bias is a risk, not a certainty. Being wise to that risk is not cynicism, it is what any statistician worth their salt strives for.

And sampling is, as we have said, inevitable: there is simply too much to count to count it properly. So let us take the extreme example of an uncountable number which we nevertheless want to know: how many fish are there in the sea?

If you believe the scientists, not enough; the seas are empty almost to the point where some fish stocks cannot replace themselves. If you believe the fishermen, there is still a viable industry. Whether that industry is allowed to continue depends on counting those fish. Since that is impossible, the only alternative is sampling.

When *More or Less* visited the daily fish market in Newlyn, Cornwall, in 2005, the consensus there was that the catch was bigger than twelve to fifteen years ago. 'We don't believe the scientific evidence,' they said.

The scientists had sampled cod fish stocks at random and weren't finding many. The fishing industry thought them stupid. One industry spokesperson told us, 'If you want to count sheep, you don't look at the land at random, you go to the field where the sheep are.' In other words, the samplers were looking in the wrong places. If they wanted to know how many fish there were, they should go where there were fish to count. When the fishermen went to those places, they found fish in abundance.

The International Council for the Exploration of the Sea recommends a quota for cod fishing in the North Sea of zero. EU fisheries ministers regularly ignore this advice during what has become an annual feud about the right levels of TACs (total allowable catches) and in 2006 the EU allowed a catch of about 26,500 tonnes.

Is the scientists' sampling right? In all probability, it is – by which we mean, of course, that it is not precisely accurate, but nor, more importantly, is it misleading. Whereas fishing boats might be sailing for longer or further to achieve the same level of catch, research surveys will not. They trawl for a standard period, usually half an hour. Nor will they upgrade their equipment, as commercial fishers do, to maximise their catch. And on the question of whether they count in the wrong places, and the superficial logic of going where the fish are, this ignores the possibility that there are more and more places where the fish are not, and is itself a bias. Going to where the fish are in order to count them is a little like drawing the picture of the donkey *after* you have pinned up the tail.

The fishermen, like all hunters, are searching for their prey. The fact that they find it and catch it tells us much more about how good they are at hunting than about how many fish there are. The scientists' samples are likely to be a far better guide to how many fish there are in the sea than the catch landed by determined and hard-working fishermen.

The kind of sampling that spots all these red herrings has to be rigorous and imaginative for every kind of bias. After all that, the sampling might still not be perfect, but it has performed a near miracle.

9
Data: Know the Unknowns

Britain's most senior civil servants, who implement and advise on policy, are often clueless about many of the basic numbers on the economy or society. We know this because we have tested them.

Numbers, if they are any use, do not fall ripe into our laps, someone has to find and fetch them; far easier, some feel, not to bother.

Much of what is known through numbers is foggy, but a fair portion of what people do not know is due to neglect, sloppiness or fear. Policies have been badly conceived, even harmful, for want of looking up the obvious numbers in the ministry next door. People have died needlessly in hospital for want of a tally to tell us too many had died already.

The deepest pitfall with numbers owes nothing to numbers themselves and much to the slack way they are treated, with carelessness all the way to contempt. But numbers, with all the caveats, are potent and persuasive, a versatile tool of understanding and argument. Treat them well and they will reward you.

They are also often all we have got, and ignoring them is a dire alternative. For those at sea with the numbers, all this should be strangely reassuring. First, it means you have illustrious company. Second, it creates the opportunity to get ahead. Simply show enough care for numbers' integrity to think them worth treating seriously, and you are well on the way to empowerment.

At frequent talks and seminars over the past ten years or more, Britain's most senior civil servants, journalists, numerous business people and academics have been asked to answer a series of multiple-choice questions about very basic facts to do with the economy and society. Some, given their status or political importance, asked to remain anonymous. It is just as well that they did.

Here is a sample of the questions, along with the answers given by a particular group of between seventy-five and a hundred senior civil servants in September 2005. It would be unfair to identify them, but reasonable to say you would certainly hope that they understood the economy. (There is some rounding so the totals do not all sum to 100 per cent, and not everyone answered all the questions.)

What share of the income tax paid in the UK is paid by the top 1 per cent of earners?

Is it …	*% who answered*
A) 5% of all tax	19
B) 8%	19
C) 11%	24
D) 14%	19
E) 17%	19

They were all wrong. The correct answer is that the top earners pay 21 per cent of all the income tax collected. It might seem unfair not to have given them the chance of getting it right, so it is reasonable to give credit to those who picked the biggest number available, 17 per cent. All the others did poorly and almost two thirds thought the answer was 11 per cent or less. Analysing the effect of the tax system, and of changes to it, should be a core function of this group, but they simply did not know who paid what. Almost as surprising as the fact that so few knew the right answer is that their answers could almost have been drawn randomly. There is no sign of a shared view among those questioned. If you were playing *Who Wants to be a Millionaire?* when this came up, you might have thought this would be a good audience to ask. You would be disappointed.

What joint income (after tax) would a childless couple need to earn in order to be in the top 10 per cent of earners?

Is it ...	*% who answered*
A) £35,000	10
B) £50,000	48
C) £65,000	21
D) £80,000	19
E) £100,000	3

The answer is £35,000. Some will resist believing that £35,000 after tax is enough to put a couple (both incomes combined – if they have two incomes) in the top 10 per cent. It is a powerful, instructive number and it is worth knowing. But the proportion of the group that did was just 10 per cent. The single most common answer (£50,000) was nearly half as high again as it should have been. And 90 per cent of a group of around seventy-five people, whose job it is to analyse our economy and help to set policy, think people earn far more than they really do, with more than 40 per cent of them ludicrously wrong.

Figure 6 **Poorer than they thought**

Actual income for top 10% compared with answers from civil servants

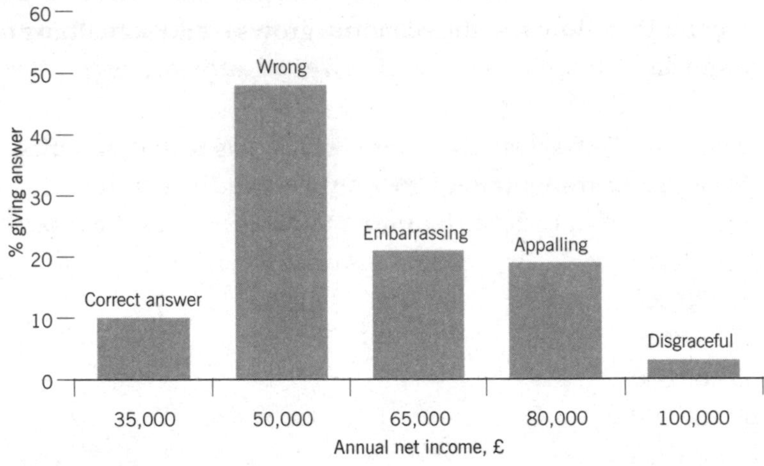

How much bigger is the UK economy now (UK national income after adjusting for inflation) than it was in 1948?

Is it …	*% who answered*
A) 50% bigger	10
B) 100%	25
C) 150%	42
D) 200%	17
E) 250%	5

The right answer is that the economy now is around 300 per cent bigger than it was in 1948. This is another question where the right answer was not an option. But since only 5 per cent of the group chose the highest possibility, it seems that most had no sense whatsoever of where the right answer lay. The economy grew in this period, on average, by about 2.5 per cent a year. More than three quarters of the group gave answers that, if correct, would have implied half as much or less, and that we were less than half as well off as we actually are. That is quite an error – economics does not get much more fundamental than how fast the economy grows – and something of a shock.

There are 780,000 single parents on means tested benefits. How many are under age 18?

Is it …	*% who answered*
A) 15,000	29
B) 30,000	21
C) 45,000	0
D) 60,000	21
E) 75,000	29

The correct figure (for 2005) was 6,000. Those who chose the lowest option once again deserve credit. But there seems to be a common belief in this group and elsewhere that we have an epidemic of single gymslip mums – a familiar political target – with half our group believing the problem at least 10 times worse than it is, and some who no doubt would have gone higher had the options allowed.

The performance on these and other multiple-choice questions over the years from all sorts of groups has been unremittingly awful. That matters: if you want even the most rudimentary economic sense of what kind of a country this is, it is hard to imagine it can be achieved without knowing, at least in vague terms, what typical incomes are. If you expect to comment on the burden of taxation, what chance a comment worth listening to if you are wholly in the dark about where the tax burden falls?

The list of excuses for ignorance is long. It is far easier to mock than search for understanding, to say numbers do not matter or they are all wrong anyway so who cares, or to say that we already know all that is important. Wherever such prejudices take root, the consequences are disastrous.

The death of Joshua Loveday, aged eighteen months, on the operating table at the Bristol Royal Infirmary began a series of investigations into what became a scandal. It was later established by an inquiry under Professor Sir Ian Kennedy that, of children operated on for certain heart conditions at Bristol, fully twice as many died as the national norm. It was described as one of the worst medical crises ever in Britain.

The facts began to come to light when an anaesthetist, Dr Steve Bolsin, arrived at Bristol from a hospital in London. Paediatric heart surgery took longer than he'd been used to; patients were a long time on heart bypass machines – with what effect he decided to find out, though he already suspected that death rates were unusually high. So he and a colleague rooted through the data where they discovered, they thought, persuasive evidence of what the medical profession calls excess mortality.

At first slow to respond, the hospital found itself – with Joshua's death the catalyst – eventually overwhelmed by suspicion and pressure for inquiries from press and public. The first of these was by an outside surgeon and cardiologist, another by the General Medical Council (the longest investigation in its history) and finally a third by the independent team led by Sir Ian, which concluded that between thirty and thirty-five children had probably died unnecessarily.

Most of those involved believed they knew the effectiveness of their procedures and believed them to be as good as anyone's. None, however, knew the numbers; none knew how their mortality rates compared.

One crucial point attracted little attention at the time – grief and anger at the raw facts understandably swept all before them – but members of the inquiry team judged that had the excess mortality been not 100 per cent (twice as bad as others), but 50 per cent, it would have been hard to say for sure that Bristol was genuinely out of line. That is, if about 15–17 babies had died unnecessarily rather than the estimated 30–35, it might have been impossible to conclude that anything was wrong. Fifty per cent worse than the norm strikes most as a shocking degree of failure, especially where failure means death. Why did mortality have to be 100 per cent worse before the inquiry team was confident of its conclusions?

The two surgeons who took the brunt of public blame for what happened argued that, even on the figures available, it was not possible to show they had performed badly (and the inquiry itself was reluctant to blame individuals rather than the system at Bristol as a whole, saying that 'The story of [the] paediatric cardiac surgical service [in] Bristol is not an account of bad people. Nor is it an account of people who did not care, nor of people who wilfully harmed patients'). Mortality one hundred per cent worse than the norm was a big difference given the numbers of children involved, large enough to constitute 'one of the worst medical crises' in Britain's history, but even then the conclusion was disputed.

The simple questions to establish the truth ought to be easy to answer: How many operations were there? How many deaths? How does that compare with others? Simple? The inquiry took three years.

Audrey Lawrence was one of the Bristol inquiry team, an expert in data quality. We asked her about the attention given to proper record keeping and the data quality for surgeon performance at Bristol. First, who kept the records?

'We found that we could get raw data for the UK cardiac surgeons register collected on forms since 1987, stored in a doctor's garage. It was nothing to do with the Department of Health, that's the point. The forms were collected centrally by one doctor, out of personal interest, and he entered the data in his own records and kept it in box files in the garage.' There was no other central source of data about cardiac operations and their outcomes.

Next, how reliable was that data?

'My own experience of gathering data in hospitals was that it was probably not going to be accurate. We were very concerned about the quality of the data. All we had were these forms, and so we went round the individual units to see what processes had been followed. We found, as we suspected, that there was considerable lack of tightness, that there was a great deal of variety in the way data was collected, and a lot of the figures were quite suspect. It was quite a low priority [for hospitals]; figures were collected in a rush to return something at the end of the day.'

So what, if any, conclusions could be drawn?

'The findings were fairly consistent that mortality in Bristol really did seem to be an outlier with excess mortality of 100 per cent, but had it been in the region of 50 per cent, the quality of the data was such that we could not have been confident that Bristol was an outlier. We were sure, but only because the numbers were so different.'

And did that conclusion mean there could be more places like Bristol with excess mortality, if only of 50 per cent, that it would be difficult to detect?

'Undoubtedly.'

It is disconcerting, to say the least, that attention to data quality can be so deficient that we still lack the simplest, most obvious, most desirable measure of healthcare quality – whether we live or die when treated – to an acceptable degree of accuracy. Why, for so long, has it been impossible to answer questions such as this? The answer, in part, is because the task is harder than expected. But it is also due to a lack of respect for data in the first place, for its intricacies and for the care needed to make sense of it. The data is often bad because the effort put into it is grudging, ill thought-through and derided as so much pen-pushing and bean-counting. It is bad, in essence, often because we make it so.

To see why data collection is so prone to misbehave, take an example of a trivial glitch in the system, brought to us by Professor David Hand of Imperial College London. An email survey of hospital doctors found that an infeasible number of them were born on 11 November 1911. What was going on?

It turned out that many could not be bothered to fill in all the boxes on the computer and had tried, where it said DoB, to hit 00 for the day, 00 for the month and 00 for the year. Wise to that possibility, the system was set up to reject it, and force them to enter something else. So they did, and hit the next available number six times: 11/11/11; hence the sobering discovery that the NHS was chock-full of doctors over the age of 90.

Try to measure something laughably elementary about people – their date of birth – and you find they are a bolshie lot: tired, irritable, lazy, resentful of silly questions, convinced that 'they' – the askers of those questions – probably know the answers already or don't really need to; inclined, in fact, to any number of other plausible and entirely normal acts of human awkwardness, any of which can throw a spanner in the works. Awareness of the frailty of numbers begins with a ready acknowledgement of the erratic ways of people.

Professor Hand said to us: 'The idealised perception of where numbers come from is that someone measures something, the figure's accurate and goes straight in the database. That is about as far from the truth as it's possible to get.'

So when forms from the 2001 Census were found in bundles in the bin, or dumped on the doormat by enumerators at the end of a bad day of hard door-knocking and four-letter hospitality, or by residents who didn't see the point; when there was an attempted sabotage of questions on religious affiliation through an email campaign to encourage the answer 'Jedi Knights' (characters from the film *Star Wars*); when some saw the whole exercise as a big-brother conspiracy against the private citizen and kept as many details secret as they could, when all this and more was revealed as a litany of scandalous shortcomings in the numbers, what, really, did we expect, when numbers are so casually devalued?

The mechanics of counting are anything but mechanical. To understand numbers in life, start with flesh and blood. It is people who count, one of whom is worried her dog needs the vet, another dreaming of his next date, and it is other people they are often counting. What numbers trip over, often as not, is the sheer, cussed awkwardness and fallibility of us all. These hazards are best known not through obscure statistical methodology, but sensibility to human nature. We should begin by tackling our own complications and frailties, and ask ourselves these simple questions: 'Who counted?' 'How did they count?' 'What am I like?'

In 2006 about £65bn pounds of government grant and business rates was distributed to local government – a large number, about £21 per person per week – overwhelmingly on the basis of population figures from the Census, for everything from social services to youth clubs, schools to refuse collection. A lot rides on an accurate Census. In 2006, preparation for the 2011 Census was already well underway (five years ahead was a trice in the planning for that Herculean task, requiring 10,000 enumerators and expected to cost about £500m, or roughly £8 a head, for every adult and child in the country). Some of the statisticians responsible for making it work held a conference about risks.

By 'risks', they meant all of us: the counters, the counted, the politicians occasionally tempted to encourage non-cooperation, the statisticians who failed – being human – to foresee every human difficulty; we are all of us just such a risk. There are technical risks too, and what we might call acts of God, like the foot and mouth outbreak in the middle of the last Census, but the human ones, in our judgement, are greatest. The whole exercise would be improved if people were less tempted to disparage data.

One of the less noticed features of the numbers flying about in public life is how many critical ones are missing, and how few are well known. One of the most important lessons for those who live in terror of numbers, fearing they know nothing, is how much they often share with those who purport to know a lot.

In the case of patient records, the flow of data creates huge scope for human problems to creep in. Each time anyone goes to hospital, there's a note of what happens. This note is translated into a code for every type of procedure. But the patient episode may not fit neatly into the available codes – people's illnesses can be messy after all: they arrive with one thing, have a complication, or arrive with many things and a choice has to be made of which goes on the form. Making sure that the forms are clear and thorough is not always a hospital priority. There are often gaps. Some clinicians help the coders decipher their notes, some don't. Some clinicians are actually hostile to the whole system. Some coders are well trained, some aren't. Although all hospitals are supposed to work to the same codes, variations creep in: essentially, they count differently. The coded data is then sent through about three layers of NHS bureaucracy before being published. It is not unusual for hospitals looking at their own data once it has been through the bureaucratic mill to say they don't recognise it.

Since Bristol, there are now more systems in place to detect wayward performance in the NHS. But are they good enough to rule out there still being centres with excess mortality that we fail to detect? No, says Audrey Lawrence, they are not.

And that is not the limit of the consequences of this difficulty with data collection in the health service. The NHS in England and Wales is introducing a system of choice for patients. In consultation with our GP, we are promised the right to decide where we want to be treated, initially from among five hospitals, and in time from any part of the health system.

Politicians have been confident that the right data can make it easier to choose where to go for the best treatment. Alan Milburn, then Secretary of State for Health, said, 'I believe open publication will not just make sure we have a more open health service, but it will help to raise standards in all parts of the NHS.' John Reid, also when Secretary of State for Health, said, 'The working people of this country *will* have choice. They *will* have quality information. They *will* have power over their future and their health whether you or I like it or not.'

The crunch comes with the phrase 'quality information'. Without quality information, meaningful choice is impossible. How do we know which is the best place to be treated? How do we know how long the wait will be? Only with comprehensive data comparing the success of one doctor or hospital with another, one waiting list with another. More often than not, this data is quantified.

At the time of writing, several years on from Mr Milburn, and after Mr Reid too has moved to another job, the best that Patient Choice can offer by way of quality information is to compare hospital car parking and canteens; on surgical performance, nothing. There is one exception, though this is not routinely part of Patient Choice. In heart surgery, surgeons have set up their own web site with a search facility covering every cardiac surgeon in the country and listing, next to a photograph, success and failure rates for all the procedures for which they are responsible (it is shortly to be amended to show success rates for the procedures they have actually carried out). Otherwise, mortality data is available for individual hospitals, but not routinely to the public through Patient Choice. It can, however, be found with persistence, and is sometimes published in the newspapers.

In Wales, it seems not to be available to the public at all. For more than a year, in conjunction with BBC Wales and the Centre for Health Economics at York, we have been trying to persuade various parts of the Welsh Health Service either to disclose how many people die in each hospital or to allow access to the hospital episode statistics to allow the calculation to be made independently. The degree of resistance to disclosing this elementary piece of information is baffling and illuminating. The Welsh Health Service argues that the data might compromise patient confidentiality, but refuses to produce even the total mortality figures for the whole country, from which the risk of any individual patient being identified is nil. So we do not know how well the whole system has been performing in even this modest respect, let alone individual hospitals. It is quite true that the data would need interpreting with care because of the high likelihood that some unique local circumstances will affect some local mortality rates, but this is not a sufficient excuse for making a state secret of them. In England, academics and the media have had access to this kind of information for twelve years. Patients in England do not seem to have suffered gross abuses of their confidentiality as a result. The Welsh authorities tell us that they are now beginning to do their own analysis of the data. Whether the public will be allowed to see it is another matter.

Not that this data would get us all that far: 'It's all right, you'll live' is a poor measure of most treatments with not much relevance, it is to be hoped, to a hip transplant, for example. Most people want a better guide to the quality of care they will receive than whether they are likely to survive it, but it would at least be a start and might, should there be serious problems, at least alert us to them.

A culture that respected data, that put proper effort into collecting and interpreting statistical information with care and honesty, that valued statistics as a route to understanding, and took pains to find out what was said by the numbers we have already got, that regarded them as something more than a political plaything, a culture like this would, in our view, be the most valuable improvement to the conduct of government and setting of policy Britain could achieve.

What can we do on our own?

There are times when we are all whistling in the dark. And sometimes, in a modest way, it works: you know more than you think you do. We have talked about cutting numbers down to size by making them personal, and checking that they seem to make human sense. A similar approach works when you need to know a number, and feel you haven't a clue.

Here is one unlikely example we have used with live audiences around the UK and on Radio 4. How many petrol stations are there in the UK?

Not many people know the answer and the temptation is to feel stumped. But making the number personal can get us remarkably close. Think of the area you live in, and in particular of an area where you know the population. For most of us that is the town or city we live in. Now think about how many petrol stations there are in that area. It is hard if you have only just moved in, but for most adults this is a fairly straightforward task, and people seem very good at it. Now divide the population by the number of petrol stations. This gives you the number of people per petrol station in your area. For us, the answer was about one petrol station for every 10,000 people. Most people give answers that lie between one for every 5,000 and one for every 15,000.

We know the total population of the UK is about 60,000,000. So we just need to divide the population by the number of people we estimate for each petrol station. With one petrol station for every 10,000 people, the answer is 6,000 petrol stations. With one in 5,000, the answer is 12,000 petrol stations. The correct answer is about 8,000. The important point is that almost everyone, just by breaking things down like this, can get an answer that is roughly right. Using the same ideas would produce roughly accurate numbers for how many schools there are, or hospitals, or doctors, or dentists, or out-of-town supermarkets.

All that is happening is that rather than being beaten by

not knowing the precise answer, we can use the information we do have to get to an answer that is roughly right, which is often all we need. As long as we know something that is relevant to the question, we should be able to have a stab at an answer. Only if the question asks about something where we have absolutely no relevant experience will we be completely stumped.

The best example of this we could come up with was 'How many penguins are there in Antarctica?' Here, it really did seem that unless you knew, you could bring very little that helped to bear on the question. Apart from penguins, you will be surprised by how much you know.

10

Shock Figures: Wayward Tee Shots

The cult of the shock figure demands amazement or alarm. A number comes along that looks bad, awe-inspiringly bad, much worse than we had thought; big, too, bigger than guessed; or radically different from all we thought we knew.

Do not succumb. When a number appears out of line with others, it tells us one of three things: (a) this is an amazing story, (b) the number is duff, (c) it has been misinterpreted. Two out of three mean they are wasting your time, because the easiest way to say something shocking with figures is to be wrong. Outliers – numbers that don't fit the mould – need especial caution: their claims are large, the stakes are high, and so the proper reaction is neither blanket scepticism, nor slack-jawed credulousness, but demand for a higher standard of proof.

> Greenhouse gases could cause global temperatures to rise by more than double the maximum warming so far considered likely by the Inter-governmental Panel on Climate Change (IPCC), according to results from the world's largest climate prediction experiment, published in the journal *Nature* this week.

These were the words of the press release that led to alarmist headlines in the British broadsheets in 2005. It continued:

> The first results from climateprediction.net, a global experiment using computing time donated by the general public, show that average temperatures could eventually rise by up to 11°C, even if carbon dioxide levels in the atmosphere are limited to twice those found before the industrial revolution. Such levels are expected to be reached around the middle of this century unless deep cuts are made in greenhouse gas emissions.
>
> Chief Scientist for climateprediction.net, David Stainforth, from Oxford University said: 'Our experiment shows that increased levels of greenhouse gases could have a much greater impact on climate than previously thought.'

There you have it: 11°C and apocalypse. No other figure was mentioned.

The experiment was designed to show climate sensitivity to a doubling of atmospheric carbon dioxide. Of the 2,000 results, each based on slightly different assumptions, about 1,000 were close to or at 3°C. Only one result was 11°C. Some results showed a fall in future temperatures. These were not reported. A BBC colleague described what happened as akin to a golfing experiment: you see where 2,000 balls land, all hit slightly differently, and arrive at a sense of what is most likely or typical; except that climateprediction.net chose to publicise a shot that landed in the car park. Of course, it is possible. So are many things. It is possible your daughter might make Pope, but we would not heed the claim, at least not until she made Cardinal. As numbers go, this was the kind that screamed to be labelled an outlier, and to have a red-flag warning attached, not an excitable press release or broadsheet headlines.

In January 2007, this time in association with the BBC, climateprediction.net ran a new series of numbers through various models and reported the results as follows: 'The UK should expect a 4°C rise in temperature by 2080 according to the most likely results of the experiment.'

That is more like it: 'The most likely results' may still be wrong, as all predictions may be wrong, but at least they represent the balance of the evidence from the experiment, not the most outlying part of it.

None of this, incidentally, implies comfort to climate-change deniers: 4°C, even 3°C, would still bring dramatic consequences, though the case would arguably be strengthened – if more modest – by using this overwhelmingly likely result (according to this experiment), rather than one that could be dismissed as scaremongering. It is always worth asking if the number we are presented with is a realistic possibility, or the Papal one.

'It could be you', say adverts for the National Lottery, which is true, though we all know what kind of truth it is. In any case, responsible news reporting owes us something better than a lottery of possibilities. And yet there is an inbuilt bias to news reporting which actually runs in favour of outliers. 'What's the best line?' every news editor asks, and every journalist knows that the bigger or more alarming the number, the more welcomed it will be by the boss. In consequence, the less likely it is that a result is true, the more likely it is to be the result that is reported. If that makes you wonder what kind of business news organisations think they are in, often the answer is the one that interests and excites readers and viewers most. It seems that consumers of news, no less than producers, like extreme possibilities more than likely ones. No wonder we are so bad at probabilities. 'Your daughter could be Pope', says the number in the headline on the news-stand. 'Gosh, really?' say readers, reaching for the price of a copy.

It is a question continually to ask of surprising numbers: is this new and different, or does the very fact that it is new and different invite caution? Examples abound. Some 18,000 years old, with the consistency of thick mashed potato or blotting paper, they turned up in a sodden cave (described by the journal *Nature* as 'a kind of lost world') and became a worldwide sensation when news of the find was reported in March 2003. The skeletal remains were soon nicknamed Hobbit Man, though the most complete of the skeletons might have been – it is still disputed – an approximately 30-year-old woman, given the name Little Lady of Flores or Flo. They were discovered in the Liang Bua cave on the island of Flores in Indonesia; hence the scientific name for what was hailed as an entirely new species of human: *Homo floresiensis*.

At about 1 metre (39 inches) tall, the 'Hobbits' were shorter than the average adult height of even the smallest modern humans, such as African pygmies (pygmies are defined as having average male adult height of 1.5m or less) and it was this that grabbed the popular imagination.

They also reportedly had strikingly long arms and a small brain. The joint Australian–Indonesian team of palaeoanthropologists had been looking for evidence of the original human migration from Asia to Australia, not for a new species. But these little people seemed to have survived until more recently than any sub-species of human apart from our own and may, some believe, be consistent with sightings of little people known locally as Ebu Gogo, reported until the nineteenth century. Some wonder if they live on in an isolated jungle in Indonesia.

They were certainly extremely unusual, apparently like no other discovered remains of any branch of the genus *Homo*: not *Homo erectus*, nor modern-day *Homo sapiens*, nor the Neanderthals. But were they a new species?

Two reasons to pause before judgement came together on 10 February 1863 in New York City – at a wedding: Charles Stratton married Lavinia Warren in a ceremony attended, as reports of the time captured that famous day, by the 'haut ton' of society. At the time, Stratton was 33 inches tall, Warren 34 inches (approximately 85cm). A few inches shorter even than the Hobbits, but intellectually unimpaired and leading a full life, they had a fine, custom-built home paid for by their many admirers and Stratton's wealth from years as a travelling exhibit for P. T. Barnum, for whom he appeared under the stage name General Tom Thumb.

The usually staid *New York Observer* reported the wedding as the event of the century, if not unparalleled in history:

> We know of no instance of the kind before where such diminutive and yet perfect specimens of humanity have been joined in wedlock. Sacred as was the place, and as should be the occasion, it was difficult to repress a smile when the Rev. Mr. Willey, of Bridgeport, said, in the ceremony – 'You take this woman,' and 'You take this man' &c.

Stratton and Warren were obviously the same species as the rest of us, perfect specimens according to the *Observer*. Their four parents and nine of their brothers and sisters were a more typical height, though the last of Lavinia's sisters, Minnie, was shorter even than she. Their existence within one species tells us, if we didn't know it already, that the human genetic programme can produce extreme variation. The world's smallest man on record was Che Mah at 2ft 2in. Robert Wadlow, the world's tallest, was 8ft 11in, more than four times as tall.

So imagine that our newlyweds had set up home on an island in Indonesia, started a family, and that their remains went undiscovered until the twenty-first century. How might we describe them? Would we recognise them for what they were, outliers in the wonderfully wide spectrum of human variation, or wonder perhaps if they too were a separate species?

The argument about whether Hobbit Man was a different kind, or simply an extreme variation of an already known kind, still rages. Most recently it has been argued that it was indeed a new species, following scans of the remains of the skull and a computer-generated image of the shape of the brain by a team at Florida State University.

The team was trying to determine whether the cave dwellers from Flores could be explained as the result of disease (microencephaly – having an exceptionally small brain – is a condition we know today), though there may have been some other, unknown syndrome which caused their particular constellation of physical characteristics. Professional rivalries and fights over access to the remains by eager researchers have complicated the investigation. 'We have no doubt it is not a microcephalic,' said Dean Falk, a palaeoanthropologist who worked with the team who discovered the remains. 'It doesn't look like a pygmy either.' The journal *Nature*, which first published news of the Hobbit discovery, headlined the latest research: 'Critics silenced by scans of Hobbit skull.'

That assessment may prove optimistic. Robert Eckhardt of Penn State University and a sceptic, reportedly said he was not convinced: 'We have some comprehensive analyses underway which I really think will resolve this question. The specimen has multiple phenomena that I would characterise as very strange oddities and probably pathologies.'

Fortunately, it is not necessary to settle the argument to make the point: unexpected and extreme findings might tell us something new and extraordinary; they might also be curious, but irrelevant to the bigger picture. The fact that they are extreme and unexpected is as much a reason for caution as excitement. Statisticians know them as 'outliers'.

Imagine that the heights of every adult human in the country were plotted on a graph. Most would bunch within 10cm or so of the average for men of about 1.75m (5ft 10in) and 1.63m (5ft 4in) for women (Health Survey for England 2004). Some, but not so many, would be further along the graph in one direction or another, at which point we might begin to call them tall or short. A very few, including Robert Wadlow, Che Mah, Lavinia Warren and Charles Stratton, would be at the far ends. The problem with the Hobbit is to decide whether it is simply a human of a kind we already know, from the extreme end of the graph, an outlier with an illness, or if it tells us that we need to create a new graph entirely, and change our assumptions about human evolution. The argument continues.

Remember our three possibilities when faced with a shock figure – an amazing story, a duff number, or a misinterpretation – and then apply them to the discovery of an adult skeleton three feet tall identified as a new species. Of the three, which is most likely?

Is the Hobbit genuinely new, different and warranting a

change to the whole map of human evolution? Or is it, like Charles Stratton perhaps, one of *Homo sapiens*' very own, but slightly erratic tee shots?

If the number is a genuine Hobbit, fine, we will have to change our thinking. But strange as Tom Thumb is, well, we knew already that the results of human reproduction can be surprising from time to time, that there will always be outliers, in all sorts of ways, but they remain, quite obviously, human. Statistics teaches us that outliers are to be expected, and so there is nothing abnormal about them, but they are certainly atypical. If that's all that was discovered in Liang Bua cave, an atypical human, it doesn't tell us anything new at all; in the scatter plot of human life, it's one wayward dot, and might be a guide to nothing more than its unusual self.

In fact, outliers are usually far less interesting than Tom Thumb. The real Charles Stratton fascinates; most statistical outliers on the other hand are not human but products of an experiment, survey or calculation, observations which are (by definition) atypical, and one of the first principles of statistics is that suspicious data or outliers may need to be rejected, not least because there's every chance that the outlier is simply wrong – measured or recorded inaccurately. Tom Thumb at least was real, even if on the fringes of the distribution. Giving credence to statistical outliers from forecasts, on the other hand, indulges fantasy.

When two years ago a company forecast UK house prices and came up with five possible scenarios, one of which was a precipitous fall in the next two years, and the other four modest rises, it would be reasonable to ask if the contrary one, so different from the others, was anything but a statistical anomaly – possible yes, but, because out of line, less probable. Guess which was reported? 'House prices could plunge', said the headlines. Funnily enough, the other four turned out to be more accurate.

The point is that outliers can be a routine fluke of the system, an erratic moment, and there will always be such blips in any distribution of height, house prices, weather forecasts or whatever. They need not be a cause for alarm. They are not necessarily a revelation.

If your business is catching drug cheats in sport, the natural variability that means some people will always be outliers is a headache. Many of the drugs that people take to make them faster or stronger are substances that are already in their system naturally. Athletes want a little more of what seems to make them good.

So cheats are detected by looking for those with extreme levels of these substances in their body. In essence, we identify the outliers and place them under suspicion. The hormone testosterone, for example, occurs naturally, and is typically found in the urine in the ratio of one part testosterone to one part epitestosterone, another hormone. The World Anti Doping Agency says there are grounds for suspicion that people have taken extra testosterone in anyone found with a ratio of 4 parts testosterone to 1 part epitestosterone. The threshold used to be 6:1, but this was considered too lax.

There are two problems. First, there have been documented cases of outlying but entirely innocent and natural testosterone to epitestosterone (T/E) ratios of 10 or 11 to 1, well above the level at which the authorities become suspicious. Second there are whole populations, notably in Asia, with a natural T/E ratio below 1:1, who can take illegal testosterone with less danger of breaching the 4:1 limit. In short, there is abundant variation.

The first person to be thrown out of the Olympics for testosterone abuse was a Japanese fencer with an astronomical T/E ratio of about 11:1. The Japanese sporting authorities were so scandalised they put him in hospital, under virtual arrest, where his diet and medication could be strictly controlled. They found that his T/E ratio remained unchanged; he was a natural outlier. It took some believing that such odd results could mean nothing, but the evidence was incontrovertible. According to Jim Ferstle, a journalist we interviewed who has followed the campaign against drug taking in sport for twenty years, the man never received so much as an apology.

Until the late 1990s this was the only test for testosterone abuse, even though we have little idea what proportion of the population would have been naturally above the 6:1 ratio which then applied, nor an accurate idea how many would be naturally above the 4:1 ratio that applies now. To add to the problem, it is well known that alcohol can temporarily raise T/E ratios, particularly among women. So there is a danger that if the outliers are simply rounded up and called cheats, there will be honest people unjustly accused.

Fortunately, there is now a second test, which detects whether the testosterone in a suspect sample is endogenous (originating within the body), or exogenous (originating outside). This test, too, has failed in experiments to catch every known case of doping (a group of students was given high levels of testosterone by a Swiss sports-medicine clinic and then tested, but not all were identified by the test as having been doped). This second test has not yet been shown falsely to accuse the innocent, but it can produce an 'inconclusive' result. One athlete, Gareth Turnbull, an Irish 1500-metre runner, told us that he had spent about 100,000 euros on lawyers defending himself against a charge of illicit testosterone use after an adverse T/E test result and an 'inconclusive' second test for the source of the testosterone, all following a night's heavy drinking. He was eventually cleared in October 2006, but not before, as he puts it, it became impossible to Google his name without coming across the word 'drugs'.

As with height and the Hobbit, we need to remember that abnormal is normal, there are always outliers, we should fully expect to see the computer spit out a figure of 11°C or higher – sometimes; but we should also acknowledge that it may tell us nothing. And if we wish to slip in a change of definition at some point – wherever that point is – at which it is decreed that normal stops and suspiciously abnormal starts, in order to label everyone beyond that point a potential cheat, or a new species, we have to be very sure indeed that redefinition is warranted. If those outliers are produced by forecasts or computer simulations, we might want to discard them altogether, or at least qualify them with the thought of the Pope taking a wayward tee shot.

Certain words invite suspicion that an outlier is in play. When we see phrases of the kind 'may reach', 'could be as high as', 'potentially affect', it's worth wondering whether this is the most likely, or the most extreme possibility (and therefore one of the most unlikely), and then to ask how far adrift it is from something more plausible. Outliers are possible, as we've seen, but don't often come to pass. It's a mildly entertaining game, each time we see the words 'could potentially affect' or similar, to supplement them with the mental parentheses: 'but probably won't'.

11

Comparison: Mind the Gap

If we compare thee favourably to a summer's day, you might accept it as a compliment, but not the basis for a league table. People and weather are categorically different, obviously. Such comparisons are impossible – with apologies to Shakespeare – without a good deal of definitional fudge. In a sonnet, we approve, and call it metaphor; but in government …

Politics has grown to love league tables and their raw material, the performance assessment: how one compares with another, who's up, who's down, who's top, who's bottom, who's good, bad or 'bog standard', who shows us 'best practice'. Comparison has become the supreme language of government. It is now, in many ways, the crux of British public policy, in the ubiquitous name of offering informed choice.

But it is more common than that. Measured comparison is the staple of political argument. Almost every 'this is better than that' attempts it. The motif to keep in mind is one everyone already knows, but has grown stale with overuse. It is as true as ever, however disguised in league tables or performance indicators, and it is this: is the comparison of like with like?

In September 2003 a teenager, Peter Williams, attacked Victor

Bates with a crowbar in the family jeweller's shop while his accomplice shot and killed Mr Bates' wife, Marian, as she shielded her daughter. Twenty days earlier Williams had been released from a young offenders' institution. Twenty months later he was convicted as an accomplice to murder. He was supposedly under a curfew order and electronically tagged at the time of the murder but, in the short time since his release, had breached the order repeatedly.

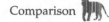

In autumn 2006 it was revealed in reports by the National Audit Office and the House of Commons Public Accounts Committee that, since 1999, convicts on tag, as they are commonly called, had committed 1,000 violent crimes and killed five people. Tagging was denounced in parts of the media as an ineffective, insecure alternative to prison, putting the public at risk and used only, it was alleged, because in comparison to prison it was cheap.

Williams had been under an intensive surveillance and supervision order. The more common use of tagging is for what are known as home detention curfews, designed for non-violent criminals and allowing them to be released from prison up to four and a half months early.

The defence of tagging was brisk, assertive and dependent entirely on spurious comparison. It was argued that of the 130,000 people who had been through the scheme, the proportion who had committed offences while tagged was around 4 per cent, comparing extremely well, said the Home Secretary and, in separate interviews, a Home Office minister and a former chief inspector of prisons, to a recidivism rate for newly released untagged prisoners of around 67 per cent. Tagging was thus a beacon of success and, while every offence caused grave concern, the scheme merited praise not censure.

Comparison is a fundamental tool of measurement and

thence judgement. If we want to know the quality of A, we compare it with B. In the case of criminal justice, the comparison is generally between the statistical consequences of the alternatives: what would have happened, how much more or less crime would there have been if, instead of B, we had tried C?

But comparison runs into all manner of booby traps, accidental and deliberate. The principal problem is well known: any comparison, to be legitimate, must be of like with like. And this particular comparison – of reoffending by convicts on tag with reoffending by others – was a model of what to watch out for, of how not to do it, or at least how not to do it honestly; an object lesson in the many ways of bogus comparison.

So let us attempt some definitional clarity. Who, when and what were the two groups being compared?

First, 'who'? The former prisoners and the tagged were not of the same kind, the tagged having been judged suitable for tagging because, prison governors decided, they were least likely to reoffend. Others were considered more risky. So was it tagging, or the selection of people to be tagged, that produced a lower offending rate? The claim of success for the scheme when the people differed so much, and in fact were chosen precisely because they differed, was, to put it politely, lacking rigour.

Second, 'when'? The period of tagging in which the new offences were committed was never more and often less than four and a half months; the period in which the Home Office counts new offences by untagged former prisoners is two years, more than five times as long. This is the second reason we might expect to see a difference in how much reoffending each group commits during the measured period, which has nothing whatsoever to do with the merits of tagging itself.

And third, 'what'? If you want to compare the success of tagging with its alternative, you have to be clear that the alternative to being tagged is to be in prison. It is not, the Home Office take note, to be at liberty. Either you are let out early wearing a tag, or you are not let out at all. These are the two groups that should be compared since these are the alternatives. The latter, being in prison, oddly enough, commits rather little crime against the public (though more against their guards and each other).

The comparison was an open and shut statistical felony, and even prompted an unusually forthright attack from the normally reserved Royal Statistical Society. Either we must compare those who were formerly tagged with those formerly prisoners, but both now free, or we must compare the different ways in which both groups are still serving sentence: the currently tagged with those currently prisoners.

Strange to tell, but the Home Office had not attempted to identify and measure recidivism specifically among people who had once been tagged but were no longer, so we had no idea, and nor did the Home Office ministers, or the former chief inspector of prisons, whether tagging worked better than prison at preventing reoffending once sentence had been served.

Tagging might well be sensible; the basis on which it was defended was anything but: a comparison of two different categories of offender across two different periods in two rather obviously contrasting places, alike only to the most superficial and thoughtless examination, where, despite all these differences, it was claimed to be the scheme that made the difference to their offending rates. And this from ministers in a department of state that also runs the police and courts and might, we hope, have a sounder grasp of what constitutes evidence.

With comparison, all the definitional snags of counting are multiplied massively, since we define afresh with every comparison we make. To repeat the well-known essence of the problem: are we comparing like with like in all respects that matter? The comparison of schools, hospitals, police forces, councils, or any of the multitudes of ranked and performance assessed ought first to be an equal race. But it rarely is, and seldom can be. Life is altogether messier than that, the differences always greater than anticipated, more numerous and significant. So we have to decide, before ignoring these differences, if we are happy with the compromise and the rough justice this implies. The exercise might still be worth it but, before making that call, it pays to understand the trade-offs. Even where intentions are good, the process is treacherous.

The big surprise of the radical shift to public policy through comparison in league tables, performance indicators and the like, an explosion of comparison unparalleled in British administrative history, has been discovering how the categories for comparison seem to multiply. Things just won't lie down and be counted under what politicians hoped would be one heading, but turn out to be complicated, manifold and infernally out of kilter. Counting in such circumstances is prone to an overwhelming doubt: what is really being counted?

For example, the government began by treating all schools as much the same thing when it put them into league tables. Today those league tables include an elaborate and, for most parents, impenetrable calculation that tries to adjust the results for every single school in the country according to the characteristics of the pupils in it. Though comparison starts by claiming to be a test of merit, it invariably collapses into a spat about underlying differences.

In 1992 school performance tables were introduced in the United Kingdom. Did the government expect still to be making fundamental revisions fifteen years later? Almost certainly it did not. In 2007 performance tables for schools faced their third substantial reform, turning upside down the ranking – and the apparent quality of education on offer – of some schools. Without any significant change in their examination results, many among the good became poor and, among the struggling, some were suddenly judged to excel. Out went the old system of measurement, in came the new. Overnight the public, which for several years had been told one thing, was told another. The government called it a refinement.

The history of school league tables (we offer a much-compressed version below) is a fifteen-year lesson in the pitiless complexity of making an apparently obvious measurement in the service of what seemed a simple political ambition: let's tell parents how their local schools compare. At least, 'simple' is how it struck most politicians at the time. One conclusion could be that governments are also prone to failures to distinguish abstraction from real life, still insisting that counting is child's play.

The first league tables in 1992 were straightforward: every school was listed together with the number of its children who passed five GCSEs at grade C or above. Though this genuinely had the merit of simplicity, it was also soon apparent that schools with a more academically able intake achieved better results, and it wasn't at all clear what, if anything, a school's place in the tables owed to its teaching quality.

For those schools held up as shining examples of the best, this glitch perhaps mattered little. For those fingered as the worst, particularly those with high numbers of special-needs pupils or pupils whose first language was not English, it felt like being obtusely condemned by official insensitivity, and was infuriating.

What's more, the results for any one school moved from year to year often with pronounced effects on the school's league-table position. Professor Harvey Goldstein, formerly of the Institute of Education, now at Bristol University, told us: 'You cannot be very precise about where a school is in a league-table or a ranking. Because you only have a relatively small number of pupils in any one year that you are using to judge a school, there is a large measure of uncertainty – what we call an uncertainty interval – surrounding any numerical estimate you might give. And it turns out that those intervals are extraordinarily large; so large, in fact, that about two thirds to three quarters of all secondary schools, if you are judging by GCSE or A-level results, cannot be separated from the overall national average. In other words there's nothing really that you can say about whether the school is above or below the average of all pupils.'

So the tables were comparing schools that were often unalike in kind, then straining the data to identify differences that might not exist. They were counting naively and comparing the tally with carelessness as to what they were really counting. Some schools, conscious of the effect of the tables on their reputation – whether deserved or not – began playing the system, picking what they considered easier subjects, avoiding mathematics and English, even avoiding pupils – if they could – whom they feared might fail, and concentrating on those who were borderline candidates, while neglecting the weakest and strongest for whom effort produced little reward in the rankings.

So the comparison, which had by now been the centrepiece of two governments' education policies, was revised, to show how much pupils in each school had improved against a benchmark of performance when they were aged 11. This was an attempt to measure how much value the school added to whatever talent the pupil arrived with. But the so-called value-added tables were nothing of the kind and unworthy of the name. (David Blunkett, who was Education Secretary at the time, described them to us as 'unsatisfactory'.) The benchmark used at age 11 was an average of all pupils at each grade. Many selective schools were able to cream off the pupils above the average of their grade and appear, once those pupils reached 16 and were measured again, to have added a huge amount of value to them. In fact the value was there from the start. These tables, misleading and misnamed as they were, were published for four years.

Then another revision was announced, this time to require school results to include mathematics and English in the subjects taken at GCSE, which in one case caused a school in east London to slump from 80 per cent of pupils achieving five passes to a success rate of 26 per cent.

Then there was a third major revision, known as contextual value added (CVA), which admitted the weaknesses of ordinary value added and aimed to address them by making allowance for all sorts of factors outside the school's control which are thought to lower performance – factors such as coming from a poorer background, a first language other than English, having special needs, being a boy, and half a dozen others. CVA also set pupils' performance against a more accurate benchmark of their own earlier ability.

In 2006, prior to the full introduction of CVA in early 2007, a sample of schools was put through the new calculation. What did the change in counting do to their position in the tables? One school, Kesteven and Grantham Girls School, went from

30th in the raw GCSE tables to 317th out of the 370 sampled. Another, St Albans C of E School in Birmingham, travelled in the opposite direction from 344th to 16th. Parents could be forgiven for wondering, in light of all this, what the comparisons of the past fifteen years had told them.

And there, so far, ends the history, but not the controversy. The CVA tables – complicated and loaded with judgements – have moved far from the early ideal of transparent accountability. It also turns out that the confidence intervals (how big the range of possible league-table positions for any school must be before we are 95 per cent sure that the correct one is in there) are still so large that we cannot really tell most of the schools apart, even though they will move round with much drama from one year to the next in the published tables. And in thinking about value added, it has dawned on (nearly) everyone that most schools are good at adding different kinds of value – some for girls, some for boys; some for high achievers, some for low; some in physics, some in English – but that the single number produced for each school can only be an average of all those differences. Few parents, however, have a child who is so average as to be 50 per cent boy and 50 per cent girl.

One significant repair looks like an accident; three wholesale reconstructions in fifteen years and you would want your money back. Unless, of course, a fair comparison was not really what they were after, but rather a simple signal to say which schools already had the most able children.

Some head teachers report great benefits from the emphasis on performance measurement brought about by league tables, particularly with the new concentration on measures of value added. They have felt encouraged to gather and study data about their pupils and use this to motivate and discuss with them how they might improve. They pay more attention, they say, to the individual's progress, and value the whole exercise highly. That must be welcome and good.

And it would absurd to argue *against* data. But it is one thing to measure, quite another to wrench the numbers to a false conclusion. Ministers often said that league tables should not be the only source of information about a school, but it is not clear in what sense they contributed *anything* to a fair comparison of school performance or teaching quality. Make a comparison blithely, too certain of its legitimacy, and we turn information into a lottery. As Einstein is often quoted as saying, 'information is not knowledge'.

Why did this happen? In essence for one reason: over-confidence about how easy it is to count. There is much in life that is only sort-of true, but numbers don't easily acknowledge sort-ofs. They are fixed and uncompromising, or at least are used that way. Never lose sight of the coarse compromise we make with life when trying to count it.

Perilous as comparison is within a single country, it pales besides international comparison. With definitions across frontiers, we really enter the swamp. Not that we would know it, from the way they are reported.

For a glimpse of what goes wrong, let us say we're comparing levels of sporting prowess; that's one thing, isn't it? And let us agree that scoring a century in county cricket shows sporting prowess. And so let us conclude that Zinedine Zidane, three times voted world footballer of the year, having failed ever to score a century in county cricket, is rubbish at sport. The form of the argument is absurd, and also routine in international comparison.

Whose health system is better, whose education? Who has

the best governance, the fewest prison escapes? Each time things are measured and compared on the same scale, it is insisted that in an important way they are the same thing; they have a health system, we have a health system, theirs is worse. They teach maths, we teach maths, but look how much better their results are. They have prisons, we have prisons, and on it goes.

Visiting Finland, Christopher Pollitt of Erasmus University in the Netherlands was surprised to discover that official records showed a category of prisons where no one ever escaped, year after year. Was this the most exceptional and effective standard of prison security? 'How on earth do you manage to have zero escapes, every year?' he asked a Finnish civil servant.

'Simple,' said the official, 'these are open prisons.'

Britain experienced a moral panic in early 2006 at the rate at which inmates were found to be strolling out of open prisons as if for a weekend ramble. By comparison, this seemed a truly astonishing performance. What was the Finnish secret?

'Open prisons? You never have anyone escape from an open prison?'

'Oh not at all! But because they are open prisons, we don't call it escape, we classify it as absent without leave.'

It is Christopher Pollitt's favourite international comparison, he says. When you reach down into the detail, he argues, there are hundreds of such glitches. Finland did not, as it happens, boast the most effective prison security in the world, nor as some might have wanted to conclude from a comparison of the numbers of 'escapes' alone, had it won, through heart-warming trust and a humane system, sublimely cooperative inmates.

At least, we don't think so, although to be honest, we are

not at all sure. Explanations are no more robust than the data they are built on. It is ludicrous, the time we devote to this, but we seem urgently to seek explanations for differences between nations – why we are good and they're bad or vice versa – when, if we cared to look, we would find reason to doubt whether the differences even existed in the terms described.

The problem begins in the simplest geographical sense: never forget the 'there' in 'how many are there?' All counting takes place somewhere, and part of the definitional headache is the need to say where. Just as when we ask how many sheep there are in a field, the field we have in mind is better if well fenced.

When it is not, we find this delicious example, in work by the Organisation of Economic Cooperation and Development – the OECD – a highly respected association of the world's developed nations with a well-regarded and capable team of researchers and economists. The OECD wanted to know – how trivially straightforward is this? – how many nurses there were in the UK, per person, compared with other countries.

'Nurse', it seems, has a settled meaning in the OECD countries; thus far, thus well defined. So a researcher contacted the Department of Health in London and asked something along the lines of: 'How many nurses do you have there?' The Department of Health answered. The OECD divided the number of nurses by the population of the UK to find the number of nurses per head.

Too bad for the OECD, health is now a devolved function in Scotland, the responsibility of the Scottish Executive in Edinburgh, not the Westminster Parliament. So the Department of Health in London defined 'there' as England and

Wales and Northern Ireland, the areas for which it still had full responsibility. The OECD used a population figure for the whole UK. How easy it is to stray from the straightest path.

It was no surprise that staffing in our health service looked anaemic. The number of nurses in England, Wales and Northern Ireland, divided by a population which also included Scotland, compared wretchedly with other developed countries.

International rankings are proliferating. We can now read how the UK compares with other countries on quality of governance, business climate, health, education, transport and innovation, to name a few, as well as more frivolous surveys like an international happiness index – 'the world grump league', as one tabloid reported it. 'Welcome', says Christopher Hood from Oxford University, who leads a research project into international comparisons, 'to Ranking World'. The number of international governance rankings, he says, has roughly doubled every decade since the 1960s.

Of course, you want to know how well Britain scores in these rankings; making such comparisons is irresistible, and even a well-informed sceptic like Christopher Hood enjoys reading them. We'll tell you the answer at the end of the chapter. First, some self-defence against the beguiling simplicity of Ranking World.

'With a header in the 27th minute followed by a second in first-half injury time, playmaker Zinedine Zidane sent shock waves through his Brazilian opponents from which they would never recover … The French fortress not only withstood a final pounding from Brazil but even slotted in another goal in the last minute.'

The words are by Fifa, the world governing body of football, describing France's victory in the final of the World Cup in

1998, as only football fans could. Two years on, the Gallic maestros, as Fifa would probably put it, stunned the world again, taking top spot in the league of best healthcare systems compiled by the World Health Organisation.

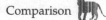

Britain finished a lowly 18th – in the WHO rankings that is, not the World Cup – a poor showing for a rich country. The United States, richest of all, was ranked 50th; humiliating, if you believed the WHO league was to be taken seriously. And though the WHO is a highly respected international organisation whose league tables are widely reported, many, particularly in the United States, did not.

The great advantage a football league has over healthcare is that in football there is broad agreement how to compile it. Winning gets points, losing doesn't, little more need be said (give or take the odd bar-room post-match inquest about goals wrongly disallowed, and other nightmare interventions by the ref). Being that easy, and with the results on television on Saturday afternoons, it is tempting to think this is what league tables are like: on your head, Zidane, ball in the back of the net, result, no problem.

But for rankings of national teams, even Fifa acknowledges the need for some judgement. For international games, each result is weighted according to eight factors: points are adjusted for teams that win against stronger rather than weaker opposition, for away games compared with home games, for the importance of the match (the World Cup counting most), for the number of goals scored and conceded. Gone is the simplicity of the domestic league. The world rankings are the result of a points system which takes all these factors and more into account, and when the tables are published, as they are quarterly, not everyone agrees that they're right. It is an example of the complexity of comparison – how good is one team, measured against another – in a case where the measurement is ostensibly easy.

Observing France's double triumph, in football and health, Andrew Street of York University and John Appleby of the King's Fund health think tank set out, tongue in cheek, to discover if there was a relationship between rankings of the best healthcare systems and Fifa rankings of the best football teams.

And they found one. The better a country is at football, the better its healthcare. Did this mean the England manager was responsible for the nation's health, or that the Secretary of State for Health should encourage GPs to prescribe more football? Not exactly: the comparison was a piece of calculated mischief designed to show up the weaknesses of the WHO rankings, and the correlation was entirely spurious.

They made it work, they freely admitted, by ignoring anything that didn't help, playing around with adjustments for population or geography until they got the result they wanted. Their point was that any ranking system, but especially one concerned with something as complicated as healthcare, includes a range of factors that can easily be juggled to get a different answer.

Some of the factors taken into account in the compilation of the WHO league are: life expectancy, infant mortality, years lived with disability, how well the system 'fosters personal respect' by preserving dignity, confidentiality, and patient involvement in healthcare choices, whether the system is 'client orientated', how equally the burden of ill health falls on people's finances, and the efficiency of healthcare spending (which involves an estimate of the best a system could do compared with what it actually achieves). Most people would say that most of these are important. But which is most important? And are there others, neglected here, that are more important?

This monstrous complexity, where each factor could be given different weight in the overall score, where much is estimated, and where it is easy to imagine the use of different factors altogether, means that we could if we wanted produce quite different rankings. So Street and Appleby decided to test the effect on the rankings of a change in assumptions. The WHO had claimed that its rankings were fairly stable under different assumptions. Street and Appleby found quite the contrary. Taking one of the trickier measures of a good health system, efficiency, they went back to the 1997 data used to calculate this, changed some specifications for what constituted efficiency and, depending which model they used, found that a variety of countries could finish top. They managed, for example, to move Malta from first to last of 191 countries. Oman ranged from 1st to 169th. France on this measure finished from 2nd to 160th, Japan from 1st to 103rd. The countries towards the bottom, however, tended to stay more or less where they were whatever the specification for efficiency.

They concluded, 'The selection of the WHO dimensions of performance and the relative weights accorded to the dimensions are highly subjective, with WHO surveying various "key informants" for their opinions. The data underpinning each dimension are of variable quality and it is particularly difficult to assess the objectivity with which the inequality measures were derived.'

In short, what constitutes a good healthcare system is in important ways a political judgement, not strictly a quantitative one. The United States does not run large parts of its health system privately in a spirit of perversity, knowing it to be a bad system compared with other countries. It does it this way because, by and large, it thinks it best. We might disagree; but to insist the US be ranked lower because of that country's choices is to sit in judgement not of its healthcare system but its political values.

It is tempting, once more, to give up on all comparisons as doomed by the infinite variety of local circumstances. But we can overdo the pessimism. The number of children per family, or the number of years in formal education, or even, at a pinch, household income, for example, are important measures of human development and we can record them just about accurately enough across most countries so that comparisons are easy and often informative. The virtue of these measures is that they are simple, counting one thing and only one thing, with little argument about definitions. Such comparisons, by and large, can be trusted to be reasonably informative, even if not absolutely accurate.

The more serious problems arise with what are known as composite indicators, such as the quality of a health system, which depend on bundling together a large number of different measures of what a health system does – how well your doctor treats you in the surgery, how long you wait, how good the treatment is in hospitals, how comfortable, accessible, expensive and so on, and where some of what we call 'good' will really mean what satisfies our political objectives. If one population wants abundant choice of treatment for patients, and another is not bothered about choice, thinks in fact that it is wasteful, which priority should be used to determine the better system?

What is important, for example, for children to learn in maths? In one 2006 ranking, Germany was ahead of the UK, in another the UK was ahead of Germany. You would expect maths scores, of all things, to be easily counted. Why the difference?

It arose simply because each test was of different kinds of mathematical ability. That tendency noted at the outset to assume that things compared must be the same implies, in this case, that the single heading 'maths' covers one indivisible subject. In fact, the British maths student turns out to be quite good at practical applications of mathematical skill – for example to decide the ticket price for an event that will cover costs and have a reasonable chance of making a profit – while German students are better at traditional maths such as fractions. Set two different tests with emphasis on one or the other but not both, and guess what happens? The reaction in Germany to their one bad performance (forget the good one) bordered panic. There was a period of soul searching at the national failure and then a revision of the whole maths curriculum.

Though the need to find like-for-like data makes comparison treacherous, there are many comparisons we make that seem to lack data altogether. The performance of the American and French economies is one example. To parody only slightly, the perception in parts of the UK seems to be that France is a country of titanic lunch breaks, an over-mighty, work-shy public sector, farmers with one cow each and the tendency to riot when anyone dares mention competition.

America by contrast, the land of turbocharged capitalism, roars ahead without holidays or sleep. And if you measure the American economic growth rate, it is, averaged over recent years, about 1 per cent higher than in France – a big difference.

Look a little more closely, though, and it turns out that the

American population has also been growing about 1 per cent faster than that of France. So it is not that the Americans work with more dynamism, just in more and more numbers. When we look at the output of each worker per hour, it turns out that the French produce more than the Americans and have done for many years – their lead here has been maintained. Even the French stock market has outperformed the American, where $1 invested thirty years ago is now worth about $36 while in France it is worth $72 (October 2006).

None of these numbers is conclusive. All could be qualified further, by noting French unemployment for example. Summary comparisons of complicated things are not possible with single numbers. When comparing such monstrously complex animals as entire economies, remember once again how hard it is to see the whole elephant.

Meaningful comparison is seldom found in single figures. Exceptions are when the figures apply to a single indicator, not a composite, when there's little dispute about the definitions, and where the data will be reasonably reliable. One such is child mortality. There is no debate about what a death is and we can define a child consistently. There will, in some countries, be difficulties collecting the data so that the figures will be approximate, as usual. But we can nevertheless effectively compare child mortality across the world, noting, for example, a rate in Singapore and Iceland of 3 children per 1,000 aged under five, and in Sierra Leone of 283 children per 1,000 (*The State of the World's Children*, UNICEF, 2006) and we can be justifiably horrified.

More complicated comparisons require a great deal more care. But if care is taken, they can be done. In Aylesbury prison in 1998 one group of prisoners was given a combination of nutritional supplements; another was given a placebo. Otherwise they carried on eating as normal. The group that received the genuine supplements showed a marked improvement in behaviour. The researchers concluded that improved nutrition had probably made the difference. Years before Jamie Oliver, the results had significant implications for criminal justice and behaviour in general, but seem to have been effectively binned by the Home Office, which refused us an explanation for its continuing unwillingness to support a follow-up trial.

Yet this comparison had merit. Care was taken to make sure the two groups were as alike as possible so that the risk of there being some lurking difference, what is sometimes known as a confounding variable, was minimised. The selected prisoners were assigned to the two groups at random, without either researchers or subjects knowing who was receiving the real supplement and who was receiving the placebo until afterwards, in order that any expectations they might have had for the experiment were not allowed to interfere with the result. This is what is known as a double-blind randomised placebo-controlled trial. Since the experiment took place in prison, the conditions could be carefully controlled.

A clear definition of how misbehaviour was to be measured was determined at the outset, and it was also measured at different degrees of severity, not just one. A fairly large number of people took part, some 400 in all, so that fluke changes in one or two prisoners were unlikely to bias the overall result. And the final difference between the two groups was large, certainly large enough to say with some confidence that it was highly unlikely to be a fluke.

This is statistics in all its sophistication, where numbers are treated with respect. The paradox is that the experiment had to be complicated in order to ensure that what was being measured was simple. They had to find a way of ruling out, as far as possible, anything else that might account for a change in behaviour. With painstaking care for what numbers can and cannot do, a clear sense of how the ordinary ups and downs of life can twist what the results seem to show us, if we are not wise to them, and a narrow, well-defined question, the researchers might just have hit on something remarkable.

As the prisons overflow and the effectiveness of strategies against reoffending seems in universal doubt, often owing to a failure to measure their effects carefully, this strategy – cheap, potentially transformative, carefully measured – remains neglected. Isn't this a little odd? Of course, there is still a possibility that the results fell victim to a rogue, confounding factor or measurement error, but the process seems to have been responsible enough. Nine years on, and the failure to pursue the findings, to try to replicate the experiment to test again whether the results were chance, is mystifying. Weak numbers and bogus numbers are hurled about with abandon in many comparisons. Here, where the numbers are strongly suggestive and responsibly used, they have been ignored.

Finally, as promised, the answer to the question of where Britain ranks internationally, taking the more serious rankings together, is, according to Christopher Hood ... in the bottom third of OECD countries; actually 11th out of 13. But you are not interested in such a Byzantine comparison any more, are you?

12

Correlation: Think Twice

This causes that. Press the remote and the channel changes. Plant the seed and it grows. When the sun shines, it is warmer. Sex makes babies.

Human (and sometimes animal) ability to see how one thing leads to another is prodigious – thank goodness, since it is vital to survival.

But it also goes badly wrong. From doing it all the time, people acquire a headstrong tendency to see it everywhere, even where it isn't. We see how one thing goes with another – and quickly conclude that it causes *the other, and never more so than when the numbers or measurements seem to agree.*

This is the oldest fallacy in the book, that correlation proves causation, and also the most obdurate. And so it has recently been observed by smart researchers that overweight people live longer than thin people, and therefore concluded that being overweight causes *longer life. Does it? We will see.*

How do we train the instinct that serves us so well most of the time for the occasions when it doesn't? Not by keeping it in check – it is genius at work – but by refusing to let it stop. Do not rest on the first culprit, the explanation nearest to hand or most in mind. Do not, like Pavlov's dogs, credit the bell with bringing the food, feel your mouth begin to water, and leave it at that. Keep the instinct for causation restless in its search for explanations and it will serve you well.

Loud music causes acne. How else to explain the dermatological disaster area witnessed in everyone wired to headphones audible at fifty paces?

That is a cheap joke. There are many possible causes of acne, even in lovers of heavy metal, the likelier culprits being teenage hormones and diet. Correlation – the apparent link between two separate things – does not prove causation: just because two things seem to go together doesn't mean one brings about the other. This shouldn't need saying, but it does, hourly.

Get this wrong – mistake correlation for causation – and we flout one of the most elementary rules of statistics. When we spot a logical fallacy of this kind lurking behind a claim, we cannot believe anyone could have fallen for it. That is, until tomorrow, when we miss precisely the same kind of fallacy and then see fit to say the claim is supported by compelling evidence. It is frighteningly easy to think in this way. Time and again someone measures a change in A, notes another in B, and declares one the mother of the other.

Try the logic of this argument: the temperature is up, the Norfolk coastline is eroding; therefore global warming is eroding the Norfolk coastline.

Or this: the temperature is up, a species of frog is dying out; therefore global warming is killing the frogs.

Or how about this: the temperature is up, there are more cases of malaria in the East African highlands, global warming is causing more malaria in the East African highlands. QED.

Convinced by these news stories from respectable broadcasters and newspapers? You shouldn't be; they are all causation/correlation errors, made harder to spot by plausibility (at least to some). Plausibility is often part of the problem, encouraging us to skip more rigorous proof and allowing the causation instinct to settle too quickly: it sounds plausible, so it must be right. Right? Wrong.

In all these cases, environmental campaigners noted that as one measurement – average global temperature – moved, so did another: the position of the coastline, the number of frogs, cases of malaria. They put these facts together and confidently assumed they had added two and two to make four and produce what they called compelling evidence, but we might prefer to call classic cases of logical hopscotch, fit for, if not derision, at least serious scepticism. All these claims have been vigorously and credibly challenged, as we shall see.

It is worth saying here and now that this chapter is no exercise in climate change denial. We need to beware another fallacy, namely that because campaigners sometimes make false claims about the effects that therefore no effects exist. That doesn't follow either. And we can note in passing that some critics of global warming have been equally guilty of spectacular numerical sophistry. The point is that even with strong cases, perhaps especially with strong cases of which people are devoutly convinced, they get carried away.

So this is a guide to a certain variety of failed reasoning, but it is a frequent failure in hard cases, given great impetus by numbers. If we state that a rare frog is dying out as a result of global warming, it sounds OK, but lacks beef; it would be more powerful if we could throw in some measurements and say that researchers believe the past decade of record temperatures has led to a 60 per cent decline in the population of red spotted tree frogs.

Put aside, if you can, your own convictions, and follow us to the point: we are concerned here only with how to avoid mistaking correlation for a causal relationship. Learn how and, whatever side you are on, you can get closer to something more important than conviction: understanding.

It is a peculiar hazard, this tendency to confuse causation and correlation, which is (a) well known and well warned against, yet (b) simultaneously repeated ad nauseam, making it tempting to say that A causes B. It is also a hazard far more widespread than debate about climate, if more easily spotted, it must be said, in examples like these:

> People with bigger hands have better reading ability; so we should introduce hand-stretching exercises in schools.

> In Scandinavia, storks are more likely to be seen on the rooftops where larger families live. Therefore storks cause babies.

Less obvious here:

> Children who come further down the birth order tend to do less well in school tests. Therefore birth order determines intelligence.

Downright controversial here:

> People with multiple sclerosis have lesions in the brain; so if we stop the lesions, we can stop the disease (i.e. the lesions cause the problem).

And here:

> Girls at single-sex schools do better than girls in mixed schools, therefore single-sex schools are better for girls.

What seems often to determine how easily we spot causation/correlation errors is how fast a better explanation comes to mind: thinking of decent alternatives slows conclusions and sows scepticism. Once again, imagination can take you far (though a thirst for more data also helps).

Good prompts to imagination are these straightforward questions: what else could be true of the group, the place, the numbers we are interested in? What other facets do they share, what else do we know that might help to explain the patterns we see? This is where the instinct for seeing causation can be put to good use, by stretching it further than the first answer to hand.

Where shall our imagination begin? With the most comical of our examples. The statement that hand size in children correlates with reading ability is true; but true because ...?

Because we generally read better as we grow up, for reasons mostly to do with maturing intelligence and education, and as we grow older our hands grow bigger. Bigger hands are a correlate of better reading, but the cause lies elsewhere; no case, then, for hand-stretching.

Next, storks and babies. This is harder, since the true explanation is less easy to guess, and there really are more storks on homes with larger families. But where does the causation truly lie? Perhaps because the house tends to be bigger as family size increases, and with more roof space ...

In both cases there is a third factor that proves to be the genuine explanation: age in the first, house size in the second.

That is one typical way for causation/correlation error to creep in. Two things change at the same time, but the reason lies in a third.

Now we begin to see how it works, what about the others, all of which have made the news?

Sufferers of multiple sclerosis have lesions in the brain. The more advanced the illness, the worse the lesions. But do the lesions cause the progressive disability characteristic of that illness? It is plausible – for many years it was thought true – and when a drug called beta interferon was discovered that seemed to arrest the lesions, it was used in the fervent hope that it would slow the disease.

There was only one way to verify the hypothesis, and that was by studying patients over many years to see how fast the illness progressed, relative to the number of lesions and the use of beta interferon. The results, when finally produced in 2005, were bitterly depressing: patients who have taken beta interferon do have fewer lesions, but are no better on average than others who have not. The progressive worsening of other symptoms seems to continue at the same rate in both groups. The lesions were found to be an effect, not a cause, of multiple sclerosis, and beta interferon was, said researchers, to use a mordant analogy, no more than a sticking plaster that failed to treat the cause.

Birth order and intelligence is also tricky. It is, once again, true that the further along the birth order you are, the worse you tend to do in IQ tests: first-borns really do perform best, second-borns next best, and so on, not every time, but more often than not, and there is a plausible explanation (watch out!) that goes like this: the more children a family has, the less parental attention each receives: the first has lots, the second maybe half as much, and so on. This is believable, but does that make it true?

Let's try the imagination test: what else could be true as you pass along the birth order? At third or fourth, or even sixth or seventh, what is plainly true is that we are now looking at a big family. What do we know about big families? One thing we know is that they tend to be of lower socio-economic status. Poorer people tend to have more children, and we also know that the children of poorer families tend, for various reasons, to do less well. So the further down the birth order you are, the more likely you are to come from a poorer family; not always, of course, but is this true often enough to be the explanation for what happens on average?

The evidence is not conclusive, but the answer is 'probably', since it also turns out, when looking at the birth order of children from the same family, that no one has found a significantly consistent pattern of performance; the last born within the same family is, as far as we know, just as likely to do best in an IQ test as the first.

The causation/correlation mistake here has been to try to explain what happens across many families (richer, smaller ones tend to do better; larger, poorer ones not so well) and claim that it applies to birth order within any one family. It looks plausible (that word again) and it seems commonly believed, but it is probably wrong.

Next, gender and school performance. It is true that girls attending single-sex schools do better academically than girls who don't. But does this prove causation, i.e. does it prove that it's single-sex education that produces the better examination results? (We'll put aside the question of what it does to their social education, which is too normative a concept to measure.) Is it, in short, the lack of boys what done it?

Again, we must use our imaginations to ask what else is true of girls in single-sex schools. We must be restless in the search for causation and not settle on the obvious correlation as our culprit. The first thing that's true is that they have relatively wealthy parents; most of these schools are fee-paying. And what do we now know from the previous example about socio-economic status and academic performance? Wealthier families tend to have, for whatever reason, academically higher performing children. Second, single-sex schools are more often selective, so that there will be a tendency to take the more able girls to begin with. So it is not surprising that single-sex schools do better: they take more able girls than other schools and these girls are usually from wealthier families. We have established that they ought to do better for all sorts of reasons, and this without taking into account any consideration of the effect of single-sex teaching.

Spare a thought for the statistician who, asked to settle this question, has to find a way of distilling school results to rid them of the effect of socio-economic background or pupil selection by ability so as to isolate the gender effect. As far as they have been able to do that, the balance of statistical opinion, once they have made these allowances, is that there is no difference.

There *is* evidence that girls tend to make slightly less inhibited choices of subject in a single-sex school, and it will almost certainly suit some pupils – which may be reasons enough for wanting your daughter to attend – but it cannot be expected to secure better exam results in general than the girls would achieve had they attended a mixed school that took pupils of similar ability.

Perhaps now, if the sensitivity of these subjects hasn't created so much hostility that we've lost our readers, we can turn to one of the most sensitive of all: climate change.

First, malaria in East Africa. It has been known for some

time that malaria in highland areas is hindered by low temperatures. These inhibit the growth of the parasites in the mosquito. The Tear Fund was one of several charities to produce evidence of an increased incidence of malaria in the East African highlands and to attribute it to climate change.

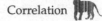

There were anecdotes: of the man from the highlands who had become landless and was living in poverty because he was bitten, got malaria, couldn't sustain work on the land, lost the land and was forced to work in bonded labour.

But when researchers looked closely at the records, they found no support for the argument. One of them, Dr David Hay, a zoologist from Oxford University, said of the records for that specific area in contrast to global averages, that: 'The climate hasn't changed, therefore it can't be responsible for changes in malaria.' His colleague David Rogers, a professor of ecology, said that some groups responded to this by accepting that there was no change in average climate but arguing that there had been a change in variability of the climate. That is an intelligent proposition, knowing, as we now do, that averages can conceal a lot of variation. So they looked – and found no significant change there either. The researchers concluded that an increase in drug resistance is a more likely explanation for the observed increase in malaria – in this instance. This was a case where there was not even a correlation at the local level, but an assumed connection between what was happening to climate globally and disease locally.

Mary Douglas, an anthropologist, wrote that people are in the habit of blaming natural disasters on things they do not like. But the loose conjunction in the back of the mind of two things both labelled 'bad' does not mean that one causes the other.

The so-called first victim of climate change was the South

American golden toad. 'It is likely,' said one campaigner, 'that the golden toad lives only in memory.'

J. Alan Pounds, from the Golden Toad Laboratory for Conservation in Costa Rica, acknowledges that the toads have been badly affected by a disease called the chytrid fungus, but argues: 'Disease is the bullet, climate change is the gun.'

In fact, the fungus does not need high temperatures, and was deadly to the toads anywhere between 4°C and 23°C. Alan Pounds is not convinced: 'We would not have proposed the hypothesis we did if there was not such a strong pattern,' he said. It is quite likely, most scientists believe, that climate change will wipe out some species. It is not at all clear in this case that it has already done so.

Climate change is confidently expected to result in a rise in sea levels. Rising sea levels may well cause coastal erosion. The temptation is to observe coastal erosion and blame climate change, as several TV news reports have done, accompanied by dramatic shots of houses teetering on clifftops.

The campaign group Friends of the Earth is more careful. Though deeply concerned about the future effects of climate change on coastal erosion, it says there has been erosion at the rate of about 1 metre a year for the last 400 years in parts of East Anglia: 'This erosion over the centuries is a result of natural processes and sea-level rise from land movements. However, in recent years the rate of erosion appears to have increased at some points along the coast. The causes are poorly understood but in addition to natural processes and sea-level rise, the effects of hard coastal defences are thought to play an important role. Ironically, our attempts to defend against sea-level rise may actually add to coastal erosion.'

Untangling climatic causation from correlation is fiendishly hard. And though climate change may make coastal erosion seriously worse in future, it is hard to claim it has made a difference yet. In fact, the rate of sea-level rise was faster in the first half of the twentieth century than in the second.

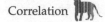

When a tit-bit of evidence seems to our taste, the temptation is to swallow it. For the non-committed too, this kind of deduction has an appeal – to laziness. It doesn't demand much thought, the nearest suspect saves time, the known villain might as well do. Even with intellectual rigour, mistakes happen, as with beta interferon. During medical trials of new drugs, it used to be customary to record anything that happened to a patient taking an experimental drug and say the drug might have caused it: 'side effects', they were called, as it was noted that someone had a headache or a runny nose and thereafter this 'side effect' was printed forever on the side of the packet. Nowadays these are referred to as 'adverse events', making it clear they might have had nothing to do with the medication.

Restlessness for the true cause is a constructive habit, an insurance against gullibility. And though correlation does not prove causation, it is often a good hint, but a hint to start asking questions, not to settle for easy answers.

There is one caveat. Here and there you will come across a tendency to dismiss almost all statistical findings as correlation-causation fallacy, a rhetorical cudgel, as one careful critic put it, to avoid believing *any* evidence. But we need to distinguish between casual associations often made for political ends and proper statistical studies. The latter come to their conclusions by trying to eliminate all the other possible causes through careful control of any trial, sample or experiment, making sure if they can that there is no bias, that samples are random when possible. The proper response is not to rubbish every statistical relationship but to distinguish between those that have taken some thought and those that were a knee-jerk.

So what, finally, about the correlation at the beginning of this chapter, of being overweight and longevity? It is true that the data from America shows overweight people living a little longer than thin people. So what is the third factor that makes unreliable a causal link between putting on weight and adding years? It is illness. Very ill people tend to be very thin. Put their fates into the mix and it has a marked effect on the results. The researchers who did the original work had to admit that they had not considered this. A statistical mistake? Yes, but more importantly an imaginative and a human one. We can all make them, and we are all capable of detecting them. Think twice.

Acknowledgements

This book began over a pizza as an idea for a radio programme that few took seriously. 'Numbers? On the radio?!' Over the years, with the encouragement and imagination of those few, it found a place on Radio 4 as the programme *More or Less*. The listeners' response was overwhelming. It acquired a growing, often devoted and delightfully interfering audience of up to a million, a web site, imitators in the press, the enthusiastic support of the Open University, the interest of publishers, a place in BBC journalist training, and finally, though we hope this is not the last of its manifestations, it became this book.

Helen Boaden was the far-sighted Radio 4 Controller who took the plunge; Nicola Meyrick the programme's incisive editor since its inception. Mark Damazer was Helen's successor and became, to our delight, *More or Less* cheerleader in chief. We owe them huge thanks.

We've been lucky to work with some talented journalists who, one by one, came through the programme as if it were a revolving door, bringing energy and a smile and leaving it, smile intact, to spread the word. Some of their reports were the basis for examples used here. Thanks to Jo Glanville, Anna Raphael, Ben Crighton, Adam Rosser, Ingrid Hassler, Sam McAlister, Mayo Ogunlabi, Jim Frank, Ruth Alexander, Paul O'Keeffe, Richard Vadon, Zillah Watson, our PAs Bernie Jeffers and Pecia Woods, and especially to Innes Bowen, whose tireless intelligence has been invaluable. Many others in Radio Current Affairs and Radio 4, studio managers, internet and unit support staff, provided creativity and quiet professionalism that enabled us to spend our time as we should, pulling our hair out over the content. Thanks too, to Gwyn Williams, Andrew Caspari, Hugh Levinson and the many colleagues and kind reviewers, the hundreds who have written to us, the hundreds of thousands who have listened, all of whom, in one way or another, have egged us on.

There have now also been hundreds of interviewees and other direct contributors to the programme, and hence to our thoughts in this book. To single out any one from the ranks of wise and willing would be unfair. All deserve our sincere thanks.

We'd particularly like to acknowledge the extensive help, comment and advice of Kevin McConway, the best kind of scrupulous and generous critic, and others at the Open University, and also Helen Joyce, Rob Eastaway, Rachel Thomas, Gwyn Bevan, Richard Hamblin, and the assistance of Catherine Barton.

Andrew Franklin at Profile Books, razor sharp as ever, and all his skilful colleagues, Ruth, Penny, Trevor and others, somehow make publishing fun and humane, even while labouring against our awkwardness. Thanks, again.

Finally, thanks to Catherine, Katey, Cait, Rosie and Julia for their love, thoughts, support and, what's most important, for being there.

We have done our best to avoid mistakes, but we know that we will have failed to catch them all, and very much look forward to hearing from readers who spot them. Thank you in advance.

Further Reading

There's a growing library of popular science books on all that's fascinating about numbers – from stories about enduring mathematical theories to histories of zero and much more – and they're often intriguing and entertaining. This list, by contrast, is of books about how to make sense of the sort of numbers that find their way into the news or are likely to confront us in everyday life. Books, in other words, to further the aim of this one.

The best of those on bad statistics is still Darrell Huff's classic, now more than fifty years old, *How to Lie with Statistics* (W. W. Norton, 1993). It is short, lively, timeless, and gives numerous amusing examples. If you were inclined to think people must have grown out of this kind of rudimentary mischief by now, you'd be wrong.

Joel Best comes at similar material as a sociologist. His concern is with why people say the daft things they do, as much as with what's daft about them, and he shows how the answer – interestingly, given the title of his books – is not only to do with duplicity. *Damned Lies and Statistics*, and now *More Damned Lies and Statistics* (both University of California Press, 2001 and 2004), have hatfuls of contemporary examples of rogue numbers, including one now famously described as the worst social statistic ever. These books are clear, thoughtful in their analysis of the way the number industry works (particularly among advocacy groups), entertaining, and offer a good guide to more critical thinking.

Innumeracy by John Allen Paulos (Penguin, 2000), is equally replete with examples, often funny, occasionally complaining, but sometimes brilliantly imaginative, about all manner of numerical garbage. It tends towards the psychology of these errors, asking why people are so susceptible to them and offers forthright answers. The innumeracy he has in his sights is partly an attitude of mind, which he strives to talk readers out of. The book also gives a useful reminder of how simple classroom maths can be used to represent the everyday world.

Gerd Gigernzer annoys one or two proper statisticians by preferring to be broadly intelligible than always technically correct. *Reckoning with Risk* (Penguin, 2003) does two valuable things well: it weans the reader off an attachment to certainty, and shows how to talk about risk in a way that makes more intuitive sense, even if it does cut one or two corners. We use the same method here.

Dicing with Death by Stephen Senn, is full of wry humour and also deals with risk and chance, particularly in health. It is more technically difficult in places, and adds chunks of historical colour, but is well worth the effort for those who want to begin to develop an academic interest.

Risk (UCL Press, 1995), by John Adams is mostly accessible, skilfully provocative on subjects the reader is surprised to find are not straightforward, and makes a sustained case about the nature of behaviour around risk. It also includes some social theory, which won't be to everyone's taste, but the power is in the numbers.

Simon Briscoe's *Britain in Numbers* (Politico's, 2005), is an extremely useful survey of the strengths and weaknesses of a wide range of economic and social indicators. He has a sharp eye for a flaw and a sardonic line in political comment.

The Tyranny of Numbers by David Boyle (Flamingo, 2001) is, as the title suggests, a polemic against the fashion for measuring everything. It overstates its case, makes hay out of the woes of certain historical figures, and altogether has a great time railing at the world's reductive excesses, just as a polemic should. For fun and provocation rather than measured argument.

Two more specialist but excellent analyses are David Hand's *Information Generation: How Data Rule Our World* (Oneworld, 2007)) and Michael Power's *The Audit Society* (OUP, 1997).

For a more formal, elementary introduction to statistics, *Statistics without Tears* by Graham Rowntree (Penguin, 1988) is a good place to start, particularly for non-mathematicians.

It's worth making one exception to our rule about books on numbers in the news, with three that are worth reading as entertaining introductions to a more numerical way of thinking, all by Rob Eastaway. *Why Do Buses Come in Threes, How Long is Piece of String* and *How to Take a Penalty* (all from Robson Books, 2005, 2007, 2003 respectively).

If you want to find out some up-to-date numbers, the website of the Office for National Statistics www.statistics.gov.uk is good for the UK, with Social Trends, www.statistics.gov.uk/socialtrends37, also a useful place to start. For an outstanding presentation of numerical information about the world try www.gapminder.org. For all manner of nation-by-nation data generally, searchable in umpteen different ways, there's www.nationmaster.com/index.php or the OECD factbook http://oberon.sourceoecd.org/vl=18803964/cl=12/nw=1/rpsv/factbook/ or you could try the United Nations Statistics division http://unstats.un.org/unsd/default.htm. Most government departments, health, the Home Office for crime, and so on, have statistics or research divisions with much useful material available online, and are generally easy to find.

There are a number of good web-based commentators. Chance News regularly steps into the fray on statistically-related news stories: http://chance.dartmouth.edu/chancewiki/index.php/Main_Page, as does STATS http://www.stats.org/. John Allen Paulos has a column which is always entertaining on the *ABC News* website: http://abcnews.go.com/Technology/WhosCounting/. Channel 4 Fact Check is usually sharp, relevant and thorough http://www.channel4.com/news/factcheck/. There are a great many other blogs and the like that regularly find their way onto statistical stories, and of varying political persuasions, but apart from http://blogs.wsj.com/numbersguy at the *Wall Street Journal*, which is consistently level headed, readers are probably best advised to find their own favourites.

Index

D

E

F

G

H

I

J